KU-437-992

SERIES EDITOR · VINCENT PAGE

DAHLIAS

PHILIP DAMP

PHOTOGRAPHS BY VINCENT PAGE

CENTURY

London Melbourne Auckland Johannesburg

The Sheffield College	Hillsborough LRC ☎ 0114 2602254
Acc. No. 488001162	
Class 635.93399 DAM	**Loan Category** ORD

WITHDRAWN

Text © Philip Damp 1987

Design: Gilvrie Misstear
Illustrations: Vana Haggerty
Production: Nick Facer

Produced by the Justin Knowles Publishing Group,
9 Colleton Crescent, Exeter EX2 4BY

All rights reserved.
No part of this publication may be reproduced
in any form or by any means without
the written permission of Justin Knowles Ltd

First published in 1987 by Century Hutchinson Ltd,
Brookmount House, 62–65 Chandos Place, Covent Garden,
London WC2N 4NW

Century Hutchinson Australia Pty Ltd,
PO Box 496, 16–22 Church Street, Hawthorn, Victoria 3122,
Australia

Century Hutchinson New Zealand Ltd,
PO Box 40–086, Glenfield, Auckland 10,
New Zealand

Century Hutchinson South Africa Pty Ltd,
PO Box 337, Bergvlei, 2012 South Africa

British Library Cataloguing in Publication Data

Damp, Philip
Dahlias.
1. Dahlias
I. Title II. Page, Vincent
636.9′3355   SB413.D13

ISBN 0-7126-0206-2

Set in Photina
Colour reproduction by Peninsular Repro, Exeter
Printed in Great Britain by
Purnell Book Production Limited
Member of the BPCC Group

CONTENTS

FOREWORD

The publication of a new book on dahlias is not a rare event, but a book so beautifully illustrated with almost fifty superb colour plates is unique. In my long association with the dahlia I cannot recall ever having seen a book on this flower so well endowed. The locations of the colour shots are well known to me: Wisley Gardens, the home of the Joint Royal Horticultural Society and National Dahlia Society annual trials, and the Show Hall at Brunel University where my own society, the Hayes Chrysanthemum and Dahlia Society, holds its September dahlia show and where photographer Vincent Page and author Philip Damp came to record some of the illustrations.

Philip Damp and I have dallied with the dahlia for almost forty years and, I have to say, enjoyed endless hours of pleasure and friendship from our hobby. His reputation as a writer is well known in the world of dahlias, and his books are read from Australia to America, from England to South Africa, anywhere, in fact, that the dahlia grows. As he has served as Secretary of the National Dahlia Society and Joint Secretary of the R.H.S./N.D.S. Dahlia Committee, his experience is second to none, and the information contained in this book bears the hallmark of that reputation. Beginners and veteran growers alike will find that they can benefit from Mr Damp's deep knowledge of his subject and perhaps have ambitions to grow the dahlia to the perfection that can be achieved.

As President of the National Dahlia Society, the world's largest specialist group for this flower, I am sure that this book will encourage such ambitions, and maybe some readers will aspire to exhibiting their blooms at the Royal Horticultural Halls in London in the not too distant future, joining the long list of skilful growers who make up the dahlia show fraternity in this country and create the world's largest and most successful show there every September. If, after reading this book, you would like to learn more about the dahlia, then allow me to extend a warm welcome to you and your friends to attend our Annual Floral Festival as it was once so-called by the founding fathers of the National Dahlia Society.

Derek Hewlett, President, National Dahlia Society

INTRODUCTION

What, you might well be excused for asking, makes a successful dahlia grower? And what distinguishes him from a gardener who loves the flower but baulks at the complications of cultivation made necessary by the demands of the show bench? There are those gardeners who devote long hours to creating perfection in any or all of the many forms that the dahlia offers and yet never take a bloom from the garden, either to grace the home or to display at a flower show. And then there is that other group, gregarious and egotistic, which grows dahlias to peaks of beauty and then unceremoniously bundles them off to the nearest flower show in search of that elusive red card of success or, joy of joys, a silver trophy!

Both groups have one thing in common – a deep and lasting love of this flower that has beguiled gardeners since it first arrived from central America almost two hundred years ago. The Victorians demonstrated this affection quite clearly. They founded special clubs and societies to foster discussion and display. Even at national level, the dahlia received the homage of our forefathers when, in 1881, the National Dahlia Society was formed, and the words "this society will popularize and extend the cultivation of the dahlia" were spoken by one of the first Presidents, Edward Mawley, at an early meeting.

Those who have succeeded these pioneers have found that they were not mistaken. There *is* great pleasure to be derived from growing dahlias as a hobby, whether you grow just for personal pleasure or whether, as so many freely admit, you embark on a prolonged ego exercise. I confess to the latter, and when my first interests were aroused in the early 1950s, it was the call of the show bench that I heard. But if you believe that this is something you can do alone, then I must disillusion you. Without the help and practical support of my wife I could never have succeeded, and my experience is one that will be confirmed by thousands of dahlia lovers worldwide. Growing dahlias for exhibition is a joint affair, and without help from your spouse it is doomed to failure. Gratefully I acknowledge the help of my wife, Liz, who has shared my dahlia life with me. In a few years time we will be celebrating our Golden Wedding, and I trust that she, too, has found the pleasure that I have.

Whether your ambitions are large or small, whether your garden is large or small, you have an opportunity to explore the intricacies of dahlia cultivation. Once you are hooked, I am told, you will always be a dahlia lover. And, like riding a bicycle or ballroom dancing, once you have done it you never forget how. May I suggest that you join us in the rewarding and so enjoyable hobby of growing dahlias? You will never regret it I can assure you.

Philip Damp, Spring 1987

The Sheffield College

Hillsborough LRC

HISTORY AND MODERN HYBRIDS

The history and evolution of the modern dahlia are very well documented, but like all good stories – and the dahlia story is a particularly interesting one – there are still many areas where speculation is rife and that seem often to owe more to the imagination of the author than to accurate historical fact.

The dahlia (a name that was not, in fact, used until the late 18th century) originated in central America where it was discovered by botanists accompanying the *Conquistadores* in the 16th century. *D. imperialis,* or the tree dahlia as it is sometimes known, bears little resemblance to today's blooms. It was called *acocotli* by the Aztecs, who used its hollow stems, sometimes 20ft (6m) or more long, as pipes to move water from mountain streams to their villages.

Although the early botanists noted the existence of the plant, it was not until 1789 that the dahlia arrived in Europe. From Mexico, Guatemala and Colombia, seeds and tubers that had been gathered over that "lost" two hundred years by Spanish colonists and were now complete hybrids found their way back to Europe. The earliest arrivals, including, it is said, *D. coccinea, D. pinnata* and *D. merckii,* formed the nucleus of the dahlia family as we know it today, although it must be added that some infusions of fresh types had been made over the two centuries.

Eventually these first basic tubers and seeds crossed the Atlantic and ended up in Madrid in the hands of a certain Abbé Cavanilles. This ecclesiastic is variously recorded as the curator or a senior officer of the Royal Botanical Gardens in Madrid, and he set about experimenting with these new arrivals. Unfortunately, he promptly lost most of them, apparently because he carefully treated them as tropical subjects and grew them in heat, whereas, except when threatened by frost, they were quite capable of growing outdoors in the more moderate climate of Europe.

Cavanilles tried again, this time with greater success, and he was assisted by one of the more important characters in the dahlia story – Dr Andreas Dahl, a Swedish botanist and a former student of Linnaeus (famous for developing the system of classifying plants). Dahl was extending his knowledge of horticulture by working on projects in Madrid with his friend Cavanilles. And a very good friend he must have been, for the Abbé gave away the right for this flower to bear his name, something that most plant hunters and scientists are loathe to do, and allowed the relatively unknown Swedish horticulturist to have the honour of giving his name to the newcomer, which was thus named *dahlia*.

Specimens of the plant, through the generosity of Cavanilles, had also found their way to northern Europe and even to Russia, where they were grown by a Russian horticulturalist, Professor Georgi, after whom they were named *georgina*. Even today, in parts of northern Europe and Russia, this flower is still called the georgina, so hard do these customs die.

Cavanilles sent seeds and tubers all over Europe first to France and then, as it became clear that this hybrid was an exciting and profitable proposition, to the rest of western Europe and, as we have seen, to Russia. In the early years of the 19th century Spanish involvement in the history of the dahlia receded, and France and Britain took centre stage. The French experimented wildly with the seeds and roots that had come to Paris and into the possession of a M. Thibaud. At *Le Jardin des Plantes* in Paris, the miserly blooms of these first dahlias, which at this stage were still single, open-centred blooms with pendant stems, did not, however, appear to be appreciated, although there is a fascinating story that the Empress Josephine favoured and fostered the dahlia. The truth is that more attention was given to the dahlia as a possible source of food. Experiments were conducted to produce massive tubers, not unlike the *pomme de terre* that had arrived from the Americas two hundred years before. However, the taste of the dahlia roots was not to the liking of French palates, and although it is recorded that tubers were used for animal feed, even cattle did not appreciate the somewhat bitter tang, and the experiment was quickly abandoned.

In Britain also food was the object of some early growers, and dahlia roots were sold in the London markets as Jerusalem artichokes. As in France, the practice was short lived. However, something more rewarding was afoot in Britain, where it was discovered that the dahlia could and would change its form, face and colour at the drop of a hat. By the early 1800s, British nurserymen were changing this sub-tropical hybrid, and the first fully double forms emerged, with a type that so fascinated 19th-century gardeners that it was cultivated almost to the exclusion of every other form. Known originally as the *globe* dahlia, because that was the satisfying shape of the flower, its name was soon changed to the more cumbersome *double show and fancy*; this title persisted until the 1950s, when it was amended to the *ball* dahlia, an Americanism adopted by the National Dahlia Society in pursuit of a better international understanding of the dahlia forms and types. The "double" of double show and fancy referred to the fact that the flower's central disc was completely covered; "show" meant a bloom of one colour, and "fancy", it follows, was a bloom of many hues.

The double show and fancy had arrived, and what a reception awaited this wonderful flower. Horticultural societies nationwide staged shows especially for the double show dahlia; nurserymen vied with one another to produce and sell hundreds of new varieties, some changing hands for as much as £500 for a single root, and there were long lists of those costing around £100 – a small fortune in those days and a situation reminiscent of the "tulip madness" that had occurred in Holland in the early 17th century.

It was not that the versatile dahlia could not produce other shapes, indeed it did regularly, but our blinkered Victorian forebears saw credit only in the double show and fancy. Even horticulturalists on the continent of Europe were obsessed with this form, but it has to be said that it was from there that the next breakthrough came. In Germany the double show and fancy was miniaturized, and the tiny, globular blooms were christened *lilliputs*, a form that we know today and that has persisted in popularity as the *pompon* dahlia. It is said that the French gave this tiny, drumstick-like dahlia its name because it resembles the pompon on the top of French sailors' hats. The nurserymen of France had imported the original pompons from Germany, but they had another form to add to the changing face of the dahlia – the *collerette*. This delicate, open-centred form has an outer ring of petals with an inner, shorter ring or collar – hence the name, which is still spelled in the French way with an "e" and not an English "a".

But it was the Dutch who have to be credited with the major break from the long-lasting confines of the double show and fancy, when in 1872 a Dutch grower produced a variety known as *D. juarezii*, which had long, thin petalling, rolled for most of its length. This was the first cactus variety, and from its progeny every cactus and semi-cactus dahlia that we have in commerce today has sprung. It is said that the Dutch nurseryman received from Mexico a packet of tubers, which were, after their long journey from the other side of the world, rather the worse for wear. They were promptly consigned to a compost heap from which, miracle of miracles, one sprang into life, was recovered and proved to be *Juarezii*. There is a contradictory story that Mexican hybridists were responsible for developing *D. juarezii*, which was, as if to confirm this story, named after a famous Mexican president. However, the Dutch origins are more widely believed and are given additional credibility by the reports of British nurserymen learning of the breakthrough and heading to the Netherlands in some haste to obtain stock and bring it back to Britain.

By now the dahlia was finding favour everywhere, and in a volatile society that was looking for the new and novel, there was a ready market for anything different. To cater for the swelling interest in Britain, the National Dahlia Society was formed in 1881. While it has to be said that the earliest members were well-to-do gentlemen who owned the land to grow flowers, the fact that they encouraged exhibitions and more exciting hybridization did lay the foundations of the flower's future, so that today every gardener has the widest choice both in terms of form and of colour. Indeed, dahlias are now found of nearly every colour of the rainbow except, unfortunately, blue. As the original hybrids were unable to produce this colour, there could not be a blue bloom. Mauves, lavenders and soft lilacs were and are still produced, some, it has to be said, very close to the sought-for true sky blue. As long ago as the 1870s, a British national newspaper offered a prize of £1,000 for the first true blue dahlia, but, sad to record, the prize was never claimed.

But the lack of one colour could not hold back the bounding progress that this immigrant from central America was making. As it continued to change its form and face, its popularity surged upwards. New forms appeared, like the strap-petalled *peony-flowered* types, which were produced around the turn of the century, and these, crossed, it is said, with the all-important *Juarezii* strain, gave us the modern *decoratives*, a group with broad, flat petals that makes up a very high percentage of the cut-flower dahlias we see in our gardens today. Other types were spawned, and their names, conjured up by both amateur and professional societies, are but memories today, names like star, mignon, aster-flowered, and decorative and cactus dwarfs.

Today the range is just as large, starting as it does with the tiniest of all the modern dahlias, the lilliputs (not to be confused with those far-off German introductions later named pompons), which are sometimes called baby or topmix types and which grow only an inch or so in diameter on plants just a foot or so high. Single dahlias (i.e., open-centred blooms) are still with us, in addition to a wide and ever widening range of named varieties, and it is easy to sow from the wide selection now listed in trade catalogues to give you a garden full of dahlias for a minimum outlay of cash.

The National Dahlia Society, in an attempt to impose order, currently has ten official groupings: *single-flowered; anemone-flowered; collerettes; water-lily forms; decoratives; ball forms; pompons; cactus-flowered; semi-cactus* and that convenient catch-all – *miscellaneous*.

Single-flowered Dahlias

If I had to give a title to the single-flowered dahlias, it would be the economy group. Single blooms, that is, those with only a single row of petals around the edge of the central disc or "eye", can be grown from seed, and although it is impossible to control the colours, a single packet costing a few pence will fill your flower bed with colour from late June until November (or the first frost, whichever comes first). Namings like "Coltness Gem" and "Redskin" (which has plants with attractive dark red or bronze foliage) produce low-growing dahlias ideally suited for bedding schemes. Parks managers love them: they fill the local public gardens with colour and delight the ratepayers – and what better combination could any manager want than a formula that ensures that his next budget goes through council unopposed!

But there are better single-flowered types in my view, and they are the many named varieties. The lilliput dahlias (sometimes known as baby dahlias or topmix, after the Dutchman, Topsvoort, who introduced them many years ago) are ideal for planting in tubs or pots on a patio or a balcony, and I have seen elegant displays of lilliput dahlias sporting themselves precariously some eight floors up on the balcony of a high-rise flat. One of the most famous of this group is the oddly-named "Inflammation", and while I was visiting Holland where it originated, I asked the raiser why on earth he had given this beautiful little dahlia such an awful name. "Ah, well," he said. "I got the name from an English/Dutch phrase book, and surely inflammation means flaming red." There is no reasonable answer to that.

Anemone-flowered Dahlias

Although they have some appeal, anemone-flowered dahlias have to be classed as one of the rarer types. Most grow only a couple of feet high, and they are better used in bedding schemes than planted as a source of cut-flower blooms.

The make-up of the anemone bloom is further proof, if proof were ever needed, of the dahlia's ability to change form dramatically. The rear petalling is composed of flat ray florets, on which is perched a sprouting mass of tiny, tubular petals. The effect is like a pincushion, but I suppose whoever gave this group its name originally might have blanched at the thought of calling his pet raising a pincushion. In truth, I have to confess a liking for these little beauties, and I hope that some enterprising hybridist will one day concentrate on them to improve the stems, which tend to be thin and whippy. There are a couple of exceptions, of course, and perhaps the best known anemone dahlia is "Comet", an attractive, dark red flower that came to us from Australia. It has excellent stems and might well be the forebear of other colours.

Collerettes

The collerette dahlias, which, as we have seen, originated in France, still maintain their place in the order of popularity, but it has to be said that they deserve much better than they get from a fickle gardening public. The collerette has everything that even the fussiest gardener might need: it has a range of colours that excludes only blue; bi-colours, tri-colours, blends and variegates abound; and its cultivation is of the simplest. The long, thin (but adequate) stems make it a joy for cutting, and the maximum size that it reaches, even when offered lots of manure, is 5–6in (13–15cm) across the face of the bloom. It has an added bonus: like daffodils and tulips in the spring, which will open in water if cut in the bud stage, the collerette dahlia will burst into bloom overnight when taken as the buds are flicking open. All in all, it is one of the best of the veteran forms, and one that has a future, even though almost one hundred years old.

Waterlily Dahlias

Possibly the most popular trade dahlias are the beautiful waterlily flowered varieties. The dahlia may be regarded as something of a horticultural chameleon so swiftly and widely does it change its form to that of other summer flowers of the British garden, and the elegant waterlily dahlias, with their broad petals and open, starlike appearance, are no exception as they correctly and in every detail emulate their aquatic namesakes.

The great strengths of this group are the strong stems that they generate and, of course, the prolific way in which they can produce blooms from July until late October or, in a mild autumn, well into November. All of them are double-flowered, and the colour range is extensive, particularly among the pastel shades so beloved by flower arrangers. In this group can be found dahlias that have served the British gardener well. Varieties include "Gerrie Hoek" (from Holland), a delicate shell-pink flower that is forty years old, and its illustrious contemporary, the yellow "Glorie Van Heemstede", which topped the Trials winners at the Wisley (1986) Royal Horticultural Society (jointly with the National Dahlia Society) Trials, winning the prestigious Stredwick Silver Medal, one of the highest dahlia awards in the world. But that is not to say that all the waterlily dahlias are old. New varieties abound, and every year new names are introduced to the lists in the whole colour range from white to purple, even approaching the elusive blue, with lovelies like the white to lavender "Porcelain", a recent introduction that took top honours on trial at Wisley.

Decoratives

The most prolific of the modern forms that we have is probably the decoratives. I say "probably", because two other groups, the cactus and semi-cactus, rival the decoratives in the range and variety of blooms produced. This great dahlia trio is outstanding because each group contains five size gradings – from the smallest (miniatures), through small, medium and large to giant, with blooms in the last category reaching diameters of 10in (25cm) or more. Of course, sizings are chiefly for exhibition use, but it is helpful if a dahlia, sold as a wizened tuber or coy little plant, is labelled with its size (as well as form) so that the gardener does not have the surprise of his or her life when it eventually flowers.

But to return to the decoratives. All are fully double, and the petals are broad and mainly involute (incurving at the margins). Some reflex to the stem, and others have petals that twist slightly as they develop, giving credibility to the unsupported story that the modern decorative dahlia is the result of a breeding between the *D. juarezii* and the peony-flowered varieties. The smaller ranges of decoratives are easily grown for garden decoration and cut-flower work, while the larger size groups happily satisfy the requirements of the exhibitor.

New varieties come onto the commercial market with great regularity and in ever-increasing numbers. The smaller ones are very easy to raise from seed (if the correct type of seed is chosen), and although it is true that the large and giant types come, in the main, from overseas, it is not unknown for some excellent giant-flowered decoratives to see light of day in this country – for example, the Surrey-based dahlia hybridist "Pi" Ensum has achieved outstanding success with recent winners such as the golden giant decorative "Hamari Gold" and the yellow large semi-cactus "Hamari Approval", both winners of the nation's top seedling trophy at the National Dahlia Society's London Show in their year of introduction.

My own preference in this section is for bi-coloured dahlias, that is, blooms in which one colour is clearly edged or tipped with another. The broad petals of the decoratives give a better canvas for such colour breaks, especially with the larger forms. The long-loved "Holland Festival", for instance, is a massive flower (sometimes 12in (30cm) or so in diameter), and yet it is a pure orange cleanly tipped with white. There has never been another like it since it was introduced in 1961, and there probably will never be.

Ball Dahlias

As we have seen, the modern ball dahlias are direct descendants of those domineering double show and fancy forms that were so beloved by the Victorians. The name ball dahlia was adopted by the National Dahlia Society and, as the word suggests, these dahlias are globular in form, with meticulously shaped petals that lie neatly, like tiles on a roof, around the whole, reflexing bloom.

There are just two size groups: miniatures (up to 4in (10cm) in bloom diameter) and the smalls, which reach to around 6in (15cm) in diameter. They are extremely popular as cut flowers and even more so as show blooms. As cut flowers their shape helps them to last (decently) in a vase for a week or more, which is longer than average, while their popularity for show-work arises from the fact that they are very easy to grow, are generally trouble-free and produce excellent stems. Ball dahlias are, in fact, ideal for the newcomer to dahlias.

Pompons

The miniaturized form of the ball dahlia is the pompon, which originated in Germany. This magical little flower performs better in sunnier regions like Australia and California, and most of our new varieties emanate from Australia where they seem to have cornered the market. The true pompon, or "drumstick" dahlia as it is sometimes called, is never more than 2in (5cm) in diameter. The neat petals reflex to the sturdy stem, and a single plant of a good pompon can produce anything between 100 and 150 blooms in one season – and in a perfect summer, it will do even better. The sheer beauty of a mature pompon bush has to be grown to be believed.

Trade groups, mainly from overseas, endeavour to sell other forms like the miniature decoratives or ball dahlias as pompons. Don't be deceived. Ask your nurseryman for the true pompon or write to the National Dahlia Society, which lists fifty of the best in its excellent booklet *The Classified Directory* (see page 94). Contrary to popular belief, the pompon does not grow on a low bush but often on a plant some 3ft (1m) high and with a girth of 6–8ft (1.8–2.4m). The plants need to be supported with canes or stakes, but any additional work required to cultivate this charming member of the dahlia family is well worth while as devotees of the group will tell you, and there are many enthusiasts who grow nothing but pompons in their dahlia plots. The Bristol and Somerset area is one such haven of the pompon, and with very few exceptions the national champions of this group have come from that district for at least two generations. They treat the pompon like philatelists treat rare stamps, corresponding with friends abroad, paying a high price for new stock and revelling in obtaining something rare and new.

Cactus-flowered and Semi-cactus

The so-named *cactus-flowered* and *semi-cactus* dahlias are the modern kinsmen of that long-forgotten raising *D. juarezii*, which, it is said, fathered so many of the dahlias we grow in our gardens today. Make no mistake, they are very popular, and although they may have originated in Europe, today they circle the globe. If you visit a dahlia trials garden, a flower show or simply the plot of a keen gardener anywhere in the world, you will find these "spiky" dahlias gracing the gardens, trials or flower shows. Japan and India have given us some elegant cactus dahlias, and from the Americas have appeared the largest of all of them, with semi-cactus forms that are reputed to grow to as much as 20in (50cm) in diameter. Not that pure size is necessarily the equivalent of pure beauty, it is not. But then beauty is in the eye of the beholder, and if you think big is beautiful then why not start with a flower?

The pure cactus form as we know it here in Britain (there are some differences internationally, but all use the same terms) offers fully double flowers where the petals are usually pointed, narrow and revolute (recurving from the margins) but can sometimes be straight or incurving at the tips. The semi-cactus flowered types, as the name suggests, are half-way between the pure or true cactus and the decoratives. The difference lies at the base of the bloom, where for the first inch or two (more on the larger size groups) the petals are broad, narrowing quickly to a pointed end.

It has to be said that for exhibition purposes the semi-cactus types are far more popular, and, if pitted against the true cactus, they usually come out on top. This is because there is more "body" or compactness in the semi-cactus flower, which appeals to judges and is more in line with the current rulings of the body that sets the standards for exhibition, the National Dahlia Society.

When it comes to cut flowers, however, the elegant, narrow-petalled true cactus types have to be preferred – they are certainly my favourites. Trade catalogues list hundreds of varieties from this group, and you can select any colour from the spectrum (apart from blue, of course). Dahlias like the soft, sulphur yellow "Klankstad Kerkrade", which was raised in Holland over thirty years ago, and the bright scarlet "Doris Day" (also from the Netherlands) adorn our gardens every summer. The millions of blooms that these plants have produced over their life span must have gladdened the hearts of millions of gardeners who, it has to be said, have come to regard them as friends.

Perhaps one of the most endearing features of the cactus form is the ease with which it copes with two, three and sometimes even more colours in one bloom. These blooms are variously called variegates, suffusions or blends (the official name), and the official listing by the National Dahlia Society quotes 50 per cent of the small-flowered group in this category. If you are looking for intriguing colours, take another look at the cactus blends.

Miscellaneous

If you thought that the range of forms and sizes that we have mentioned so far was enough for any species, you would be wrong. The dahlia has a host of other forms, some of which are not officially recognized here in Britain and some which are.

Named varieties of *dwarf bedders*, for example, are extremely popular, their name giving the clue to their strength. Raised by taking cuttings or dividing tubers (see pages 21–4), they will ensure that whatever colour pattern you want for your flower bed is provided.

Orchid and *double-orchid flowered* dahlias, with their delicate, orchid-like form, are the favourites of flower arrangers, but not, apparently, of the Dahlia Society officials, who lump them into the category that, for want of a better word, they call miscellaneous.

Then there are the *chrysanthemum-flowered* dahlias – enough to make a chrysanthemum enthusiast wince – which have large, shaggy, incurving and recurving bloom-heads, just like the queen of the autumn. They were raised in Holland but have never caught on in Britain. A similar fate has befallen the *rose-flowered* dahlia. The Japanese introduced the first of these, the exotically named, bright red "Kuroi Hitomi", which, when in the bud stage, is remarkably like the form of an opening rose. Again, there was no clamour for more. Perhaps modern gardeners are rather more spoiled for choice than those who were around in the

19th century. They would have trampled over one another to get tubers of such a variation.

Finally, mention must be made of the *carnation-flowered* types, which did become popular at one period. Grown in a range of colours in imitation of the favourite wedding flower, they were basically a miniature decorative in form, but with the petal ends on every bloom split permanently into two or three points. The overall effect is that of a rather fluffy flower, which in Britain is correctly called *fimbriated.* In the United States these dahlias are referred to as *laciniated* types, but the best description is the French *dentelle,* meaning lace. Thus we have the well-known "Dentelle de Venise" ("Venetian Lace") or the scarlet "Dentelle de Velours" ("Velvet Lace").

So it goes on with this versatile flower, because this is not the end of the story. Dahlia enthusiasts all over the world know that there are other forms to be realized: there is still a blue dahlia to discover and one that has perfume to enhance its lovely face. Perhaps there is the chance that a group of frost-resistant dahlias may be found. Imagine a blue dahlia with a fascinating perfume that could withstand the rigours of a British winter.

No one, it seems to me, has ever been able to control or collate the dahlia. Old forms, the star and mignon types, for example, came and disappeared. But can a dahlia once established disappear? It may well do so in the form in which it first appeared to the horticultural world, but it cannot go away altogether. One day it will return, possibly in the same guise, possibly with some variations. It is this uncertainty and the promise of the "pearl in the oyster" that has caused so many skilled hybridists to devote their lives to the cultivation of *D. variabilis* and hope that they might be the one to open new avenues of discovery.

CULTIVATION

Of all the summer plants that we have in cultivation today, the dahlia is one of the easiest to grow. By the same token, if a little care and attention are given to the needs of this versatile species, the results that you achieve might well amaze your friends, your neighbours and perhaps even yourself. Although it will grow in virtually any type of soil, the dahlia does have a few pet hates that the first-time grower should endeavour to avoid. First, the dahlia does not like badly drained situations; if your garden is liable to flooding or becomes very heavy after continuous rain, try raising the dahlia bed slightly – say about a spade's depth – by piling the soil forward as you dig. This minor assistance will pay dividends as the vigorous root system thrives in the open medium you will have created.

Situation is another important aspect of successful dahlia cultivation. Avoid choosing a site that is close to high walls or shrubs or under a tree. Such a position will only "draw" the plants, thinning the elongated stems and, of course, reducing the quality and quantity of the blooms. If you have a choice, pick an open, well-drained spot that gets sunshine for the greater part of the day but that receives some respite (from lengthening shadows perhaps) later in the evening. "Spot" planting in, say, a herbaceous border is not advisable as the dahlia always seems to resent the competition when it has to fight for root-room with other summer flowers. A bed or small corner of the garden entirely devoted to your dahlias is the most satisfactory solution, and if you are able to spare the room to do this for your dahlias, they will get off to a flying start and will probably be the most successful dahlias you have ever grown.

Nature has provided the dahlia with a superb mechanism for producing plants. Every dahlia plant, even those grown from seed in their first year, will have a large root or tuber (as it is better known) after a full season in your garden. This root, when lifted and stored (see pages 30–2) will offer you two of the three methods of propagating dahlias – it will provide cuttings that can be rooted independently from the tuber and it may be divided into two, three or even more portions so that several plants will grow where just one grew before. The third method of propagation, is, of course, by the sowing each spring of seed that has been harvested from plants grown in the previous year (see page 24). But firstly, let us look more closely at those generous tubers.

If you are considering raising your own cuttings or tuber divisions, you will need some equipment. It is possible to root cuttings and divide tubers using just a coldframe, but for best results you will need a small greenhouse that has enough heat from the end of February to exclude frosts at night and maintain at all times a good growing temperature of a minimum of 50°F (10°C). Heat, as we all know, is not cheap, whatever source is chosen, so it may well be that you would prefer to buy your plants later on, in late spring, perhaps, when all fear of frosts is over, so that you can set them out in the open garden as soon as they are received. What you would miss, if you chose to do this, is the thrill that every gardener finds when he is able to create his own plants. And there is no greater satisfaction than the actual creation of living roots on a once-naked stem, an achievement that even horny-handed veterans still marvel at.

Propagating Dahlias

Taking cuttings   Once you have decided that you want to do it all yourself, take your favourite dahlia tubers into the greenhouse in late February and turn up the heat. Prepare boxes or trays of a sandy loam or a soil-less compost and push the tubers into the compost. The tubers should be firmed into the surface of the compost, rather than buried or planted, because the new growth that will give us our plants or divisions arises from the crown of the tuber, and that is where the old tuber stem and the fatter portions of the roots meet. New growth occasionally

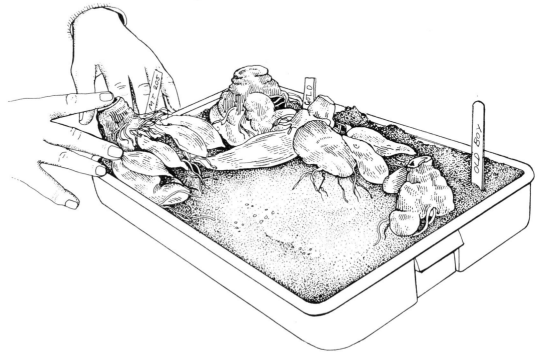

When you have trimmed the old tubers, set them in shallow trays of peat. Placed in a coldframe they will soon offer growth "eyes", allowing you to divide each into several pieces.

21

comes from further up the stem at the old leaf axils, but the strongest and best shoots come from the area of the crown. It is essential that this valuable crown position is free from possible rot or deterioration, and pushing the tuber into the compost allows you to position the growth area or crown above the moisture level and away from danger.

To induce growth water the compost well and turn on the heat to make an irresistible combination that convinces the dahlia tuber that summer has arrived so that it starts to produce new growth in anticipation.

If your dahlia roots are healthy, the first growth eyes or buds will appear on the crown within a couple of weeks. Two weeks more, and the shoots will be 3–4in (7–10cm) long, quite long enough to be removed from the parent and start a new life of their own. But to obtain those shiny new roots that are going to develop into beautiful plants you will need a little more equipment.

The successful rooting of young dahlia shoots needs a closer atmosphere and higher temperatures than those that you provided for the sprouting tubers in the trays. Ideally you need a small propagating case. These come in all shapes and sizes and are widely available in our garden centres and supermarkets. Pick one that meets your needs: one that will hold a hundred cuttings or more might well be far too large, and you would be better off with a small tray covered with a plastic dome. Some of the propagators manufactured nowadays have built-in heat, with soil-warming cables threaded through the base and a thermostat to control the rooting temperature, which should be around 70°F (20°C). The

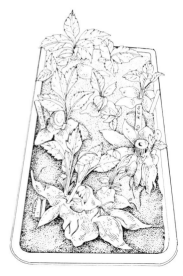

Dahlia tubers sprouting excellent cuttings that can be taken from the old tubers and rooted to make new plants.

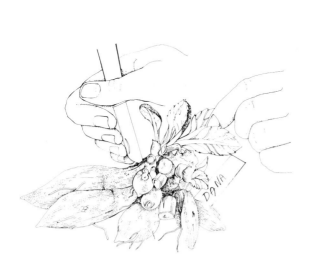

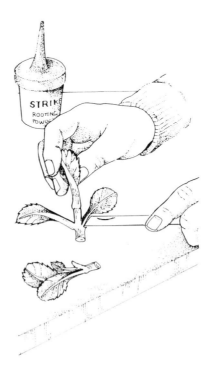

Sever cuttings from the parent plant with a sharp knife (far left) and trim the cuttings (left) before rooting them.

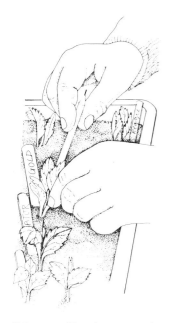

Set your cuttings in a propagator.

wonders of modern science have completely done away with the need for green fingers.

Fill the tray or multiples of trays in your propagator with a good rooting compost. You will find many on the market, but my own formula is a fifty/fifty mixture (by bulk) of peat and coarse sand. Make sure that the peat is moist, and, for the best results, rub the two constituents together by hand so that you hear a satisfying "rasping" sound. Firm the medium you have prepared into the trays so that it is ready to receive those cuttings.

Use a sharp knife or razor blade to take the cuttings. If possible, sever them from the parent root as close as possible to where they join the crown. Trim each cutting below a node (the point where the young leaves join the stem) and dip them into a hormone rooting powder, which will cauterize or seal the wound you have made with your knife. Now push the prepared cuttings into the trays of rooting mixture to a depth of approximately 1in (25mm) and cover them with the plastic dome. It is now up to you to ensure that the minimum rooting temperature is maintained at 70°F (20°C), and on sunny or even very bright days, shade the cuttings with a double thickness of newspaper.

It takes around two weeks for the cuttings to develop new roots, and you will know when this miracle of nature occurs. The cuttings will stand up perkily, and, when tugged gently, will cling tenaciously to the rooting medium, telling you that it is time for them to be moved on. The little plants (because that is what they are now they have their own roots) need something substantial in the way of nutrients to keep them growing, for the peat/sand mix in the rooting trays is absolutely sterile. Carefully extract the cuttings from the propagator and plant them up in trays or, perhaps better, individual pots using John Innes No. 1 potting compost or, the most popular choice these days, a soil-less compost like Levington or Arthur Bowers mixture.

The plantlets are now able to look after themselves and will thrive, putting on a great deal of leaf and stem in the next month, during which time you might consider moving them into a coldframe, where they can develop until the end of May when they can be set out in the chosen flowering position in the open garden.

Dividing tubers If, rather than take cuttings, you prefer to propagate your dahlias by dividing the tubers, the initial steps in the process are the same. Start the tubers as if you were planning to take cuttings, but when the shoots on the crown are only approximately 1in (25mm) high, take the whole root from the sprouting tray and place it on a firm bed on the staging. The object is to cut the whole root into as many pieces as possible, remembering that each piece must possess at least one "eye" or growth point, that is, some of the old tuber and a portion of the new roots that will have grown in the tray. The first cut can be made with a sharp knife or small saw – I find that an old bread-knife is perfect – and, aiming down the stem between the two most prominent shoots, make the

After two or three weeks, remove the rooted cuttings from the propagator.

first split. Now look carefully at the two pieces and assess whether further cuts can be made to provide more plants. It is often possible to get four or five divisions with little difficulty.

Once the whole root is in pieces, you should pot them up. This time use deeper trays of good John Innes No. 1 or Levington compost and let them grow on until you need them. Like the rooted cuttings, they may be moved into a coldframe to await planting, which should be scheduled for the first occasion when the spring frosts are over. Setting a dahlia plant in full leaf in the open garden when frost is prevalent would be tantamount to horticultural homicide!

Sowing seeds Late March is soon enough for seed-sown plants, and you can raise them on your windowsill if you have no other place. The resulting plants are, of course, just as susceptible to frost damage as those raised from cuttings or divisions, but if you take care, they will make lusty plants for the garden later on. It has to be said that the raising of seed-sown dahlia plants cannot offer such exciting ranges of form and colour as those propagated from cuttings, as most will be low-growing, open-centred types in a variety of colours. Many gardeners like this uncertainty, and there is no doubt that the major seed firms have made such progress with dahlia seed that they can now offer some exotic ranges, including even the possibility of double-flowered types in a variety of forms. Indeed, some have taken international awards in European seed trials and are very good. But they will never offer you the choice that the tubers do. If you really want to re-create that exquisite scarlet pompon or the eye-searing yellow cactus that you noticed at your local show last year, you will have to find out the variety name and use cuttings or divisions from it to satisfy your needs.

There is one aspect of growing dahlias from seed that does offer some real excitement. That is the creation of new varieties by using the seed saved from selected stock from one season, to plant again the next in the hope of finding a good new variety. Those that choose to pursue this facet of the dahlia hobby are few and far between, because they will sow hundreds, possibly even thousands, of young dahlia seedlings in the hope, often a vain one, of finding a really marvellous new type. What most such dahlia growers will always have is a garden full of hybrid dahlias in all sorts of forms, a rainbow of colours and heights that vary from mini to maxi with alarming regularity. The stems will be whippy and suspect, bend in the slightest breeze and allow the flowers to arc over in a most unbecoming manner. I suppose this may have a certain attraction but a garden full of the classical forms and colours, controlled by tuber and plant, is the choice of the majority of dahlia enthusiasts, and that must count for something.

There are, of course, the gardeners who will have none of this. Away, they say, with the messy potting, the cold nights spent popping in and out to the greenhouse and the devil-come-frost that undoes all that patient work in one fell

Split a tuber to obtain several divisions.

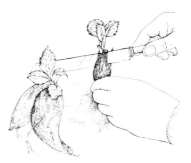

A tuber successfully divided into a number of parts.

swoop. Their solution is to comb the garden centres and nurseries in the spring, searching, and ready to part with hard cash, for the plants that have been raised by the grateful nurseryman. It would be an odd world if we were all the same, so although I would say that they are missing a lot of the fun of gardening, they certainly do contribute to the dahlia scene, if only because of their demands for perfection. If you are raising your own plants, you will, from time to time, put up with a slightly inferior plant – the odd leaf missing here and there or a crooked stem do not seem to matter; after all, it is your very own plant. But if you are buying plants, you look for completely flawless examples. You would be silly if you did not; after all, it is your money. So the gardener in the nursery or garden centre is placed on his mettle: if he raises inferior plants, they will still be on his hands at the end of the spring selling period.

The gardener searching for the best dahlia plants must know exactly what to look for. First, avoid diseased plants. A dahlia plant infected with virus will never get healthy again; there is no cure, and any plant that is stunted, has disfigured foliage and, the most telling identification of all, has pale green and yellow marks on the veins of the leaves, should be left on the nursery shelf. Small, sad plants would never catch your eye anyway, but beware of the opposite, long, drawn plants, with excessive distances between the leaf joints. Any plants that have been left to dry out and have hardened at the stem base, which will start to look like tree bark, should also be avoided. The perfect specimens are those that have dark, glossy green leaves, a leaf-spread equal to the height of the plant and thick, sturdy stems.

If you prefer, tubers bought early in the year can be held back for planting out. Nowadays, you can purchase a wide and exciting range of dahlias in pot-tuber form, and these will give you excellent results. Although the expression "pot-tuber" may be new to you, pot-tubers are certainly not new to the dahlia trade, which produces them by the million every year for sale, growers in the Netherlands alone exporting 50 million in 1985. The pot-tubers that you see on sale will have been grown within the confines of a pot (hence the name) during the whole of the previous summer, and the pot will have restricted the size. What you actually buy is something resembling a cluster of grapes or ping-pong balls, carefully encased in a plastic bag and fronted with a colour shot of the end product. Such dahlias are the stuff of which the modern plants are made. Do try them, because, despite their size, you will get magnificent plants in the same season, each tuber producing dozens of blooms for you from July to the first frost. In addition, because they sell so well, the price of the pot-tubers is right. In the spring of 1986 a single, named tuber of a pompon, cactus or decorative pot-tuber cost between 60p and 80p, which is an absolute bargain when you consider that, with care, you can keep the variety going for years. Some of my friends in the dahlia world have stocks of old favourites that go back for thirty years, and I recently grew the stock of a dahlia named "Kaiser Wilhelm", which had been given to me by a German friend – it was 100 years old! But, if you do avail yourself of such dahlia bargains, how do you go about growing them?

Planting Out

Pot-tubers, in fact any tubers, do not have to be subjected to greenhouse treatment to make them grow. In most parts of the United Kingdom they can be planted, unsprouted – that is, showing no growth eyes – in the place you wish them to flower. Excavate a fairly large hole, say 1ft (30cm) deep, and half-fill it with peat, well-rotted compost or (if you are feeling generous) soil-less compost. Set the tubers onto this pleasant starting medium and fill the hole to soil-level with the same mixture. If you do this at the beginning of May, the new growth will not show itself above the soil level until the end of that month, by when you should be safe from frost. Should the shoots emerge earlier and frost is forecast, it is a fairly simple task to "earth-up" (cover with soil) the dahlias as you would with potatoes or to cover the early greenery with a large pot or cardboard box. Once the tubers have started to grow there is virtually no stopping them, and by July you will have a massive bush with strong stems carrying buds and showing a lot of promise for a colourful summer.

As we have seen, the planting out of rooted cuttings takes place in May, although it may be a little later in the north of England and Scotland, but it is always dependent on the presence or otherwise of frost. In late May and early June you will find dahlia growers anxiously scanning the evening sky as the clouds clear and a frosty night looks to be in the offing. It is then that they scrutinize the weather forecasts or ring up the local weather station. When the all-clear does sound, the young dahlias can be set out and the positions marked with a stout cane. The distances between plants are important. Although the small, bedding dahlias can be set fairly close together so that they blend in with one another, it is advisable to leave about 2ft (60cm) between the taller growing, named varieties. The single cane that you use to mark the position can be supplemented later on by two other canes, which, when placed in the form of a triangle, will allow the plants to grow upwards and outwards but yet be contained efficiently by strong fillis or garden twine within the three canes. Encourage young plants to form a bush by removing the central growing point very early. The first sign that the young dahlia is maturing may well be the appearance of a small bud in that growing point. Do not let this flower; in forming a bloom, which will always be a sub-standard one, it will interfere with the forward progress of the plant and delay the formation of the already forming bloom-bearing laterals. Remove such early bud appearances, and within a few days you will notice that those bloom stems nestling in the leaf axils have fattened and are thrusting upwards to give you your first flowers of the summer.

Take out the healthy plants from their pots.

Set out the dahlia plants in the prepared bed.

After planting, the young dahlias should be hoed in.

Encourage side growths by stopping a plant.

Caring for your Dahlia Blooms

By the end of June, certainly by early July, the bedding varieties will be in bloom. To keep them growing and blooming consistently, you must remove the dead and dying flowers regularly. The taller growing varieties, especially those grown from tubers or divisions of tubers, will be in bloom during July, certainly no later than the first weeks of August, even in a bad summer. As the buds rise and open, remember to keep the stems well tied into the cane triangle, and, if you would like better quality blooms than those that the plant will produce naturally, you might like to practise a little judicious disbudding. Disbudding is something that many gardeners dislike; they say it is unnatural and that nature should be allowed to do its own thing. Well, the purists may have a point, but removing a few surplus buds that are in any case replaced by the dahlia within a few weeks, will give you larger and brighter blooms and longer and stronger stems that are ideal for cutting for home arrangements and will not harm your plants in the slightest.

To achieve these higher quality blooms, wait until the first buds appear at the end of each stem. Almost without exception there will be three buds in a smaller cluster: a main (or larger) bud with two ancillary buds alongside. Remove the ancillaries, leaving the larger bud to bloom, and this will immediately give you a longer stem. Take away the small buds in the first pair of leaf joints below the cluster, and that stem can be 2ft (60cm) long. No damage is done, and the removal of this small surplus of bud will allow the new growth further down the plant to accelerate and lengthen, soon replacing the blooms that you cut.

Blooms will continue to appear until the first frosts, but, as with the bedding varieties, you must remove the dead and dying blooms to maintain continuity. Nothing is more certain to slow down a dahlia plant than the leaving of dead flowers; seedheads form at the expense of growth, and the loss of blooms is dramatic.

Apart from taking resolute steps to combat pests and diseases (see pages 28–30), you must ensure that you provide your dahlias with two additional requirements: water and ample food.

Water constitutes 95 per cent of the dahlia's structure, and good blooms are acquired only by ensuring that the plants never want for this very special commodity. A wet season is, it follows, good for dahlias, but even when nature is bountiful, there will be occasions during the summer months when they need you to help and when the hose or watering can is your best friend. Dahlia enthusiasts know that plants should never be allowed to dry out around the roots, and a mulch of straw, compost or well-rotted farmyard or stable manure laid around the dahlia plot at the end of June or in early July will conserve moisture very effectively. If the soil surface is moist, the powerful young feeder roots will be able to spread and probe, increasing the amount of moisture and nutrient available to the plant and, of course, giving you more blooms, a longer flowering period and strong, healthy plants.

The additional food can come in one of several ways. For example, when you mulch your plants, a certain amount of goodness is leached into the soil from your mulching materials, and if you are able to use rich, well-made compost or stable manures, then this might well be all that you need. Where such manure is not available, you should enrich the soil in which your plants will grow before planting. Autumn digging can be supplemented by adding whatever the garden has to offer at that time of the year – garden bric-à-brac, leaves (oak and beech are best), the used contents of grow-bags, hop manures (available from many garden centres), peat, in fact anything that will eventually rot down and help the normal processes of nature. In the spring, fork the plot over, layering any of the un-rotted leaves or manures into the bottom of the trench. When you have finished, top-dress – that is, scatter around on the soil surface – a fine-grade bone meal fertilizer at the rate of 4–6oz per sq yard (110–160g per sq metre). Lightly hoe or fork this into the surface, and if you do this about a month before planting – say in late April or early May – your soil will be in great shape to produce first-class dahlias.

If you want to give an additional boost to your growing programme and ensure an abundance of bloom, summer feeding is recommended. Nowadays, everyone summer feeds their garden because it is so easy to do and takes no time at all. There are several ways to add summer goodness: by watering liquid feed via your watering can or hose directly to the roots of the plants; by foliar feeding, that is spraying the same liquid feed onto the leaves for almost immediate absorption; or by using granular feed that is scattered onto the soil surface and taken in by rain or your watering can. All are effective and are best applied in the evening. Never, never exceed the recommended measures that you will see on the packet or bottle, and always follow the manufacturer's instructions. It is best to use less than the maximum dose, but water it in at regular intervals, and you may like to choose one day, Sunday, for example, which becomes "feed day" to remind you to carry out the work. You must make your own choice from the many proprietary brands on the market, but in June, July and August use one that is high in nitrogen, changing to a high-potash feed in September and October. A weekly programme is recommended for July and August, with fortnightly gaps for the remainder of the season.

Pests and Diseases

Throughout the growing and blooming period, there are several jobs that you must do to maintain the health and vitality of your dahlias, and the first is to pay attention to hygiene. Like all the other flowers in our summer gardens, the dahlia receives the unwanted attentions of a host of predators. From the moment the tubers start into growth for cuttings in February until the frosts of autumn end the season, there will always be something trying to thwart your efforts. To ignore such attacks and do nothing will result in a drop in quality that may, on

Apply a mulch of farmyard manure or straw to help conserve moisture and keep down weeds.

Give a liquid fertilizer during the summer.

occasion, startle you. Attacks from slugs and snails in the warm, spring evenings, for example, can completely strip your young dahlia plants; similarly, when the insidious earwig breeds, swarms of young appear in June and August, and the earwig armies can tear both plant and buds to shreds. You must counter-attack, and not just by making a token gesture. You must adopt a well-planned programme that not only rids the plants of pests *in situ*, but incorporates such deterrents that even the bravest invaders will not venture onto your plot.

To ensure that your dahlias remain clean and growing healthily, you must adopt a strict regime. From the moment that your plants are rooted in the greenhouse, through their brief life in the coldframe and then into your garden, you should maintain a weekly programme of spraying. The market is brimming with good insecticides nowadays, and my advice is to use one based on malathion. Once each month, use an alternative like permethrin; this will help prevent your unwelcome guests building up immunity, which they can easily do if treated with only one type of spray. These suggested sprays, applied both over and under the foliage, will control most of the aphids – green- and blackfly and the whole family capsid – which all contribute to the possibility of your dahlias contracting one of the diseases, like the incurable mosaic virus, which are transmitted by insects moving infected sap from one plant to another.

Those nocturnal invaders, slugs and snails, are a different proposition. Fortunately, they are slow moving, and there are so many methods of control available these days that you should not see much of them if you take strong steps to assert your superiority. Put down a bait in the form of small bran pellets that have been impregnated with the chemical metaldehyde, which is irresistible, but fatal, to slugs. Scatter the pellets evenly among the plants at planting time and also when the dahlias are in the coldframe (when they are especially vulnerable). Clearing up the victims can be a little messy but may give some satisfaction.

Perhaps the most persistent predators are earwigs and caterpillars. Your regular, weekly spraying programme will deter them, but should a change in the weather (both like warm temperatures) increase the attacks, you can trap earwigs by placing clay pots (plastic ones tend to disappear in the breeze) on the top of your canes or stakes. The night-feeding earwigs have the strange habit of seeking the dark recesses of such pots in the morning, and all you will have to do is empty the contents into a small container of paraffin or strong insecticide to end their marauding days forever. The earwigs' second breeding cycle takes place in August (the first is usually in June), and the young earwig nymphs can be a nuisance as they feed on opening buds and blooms. Here you can take a leaf from the exhibitors' methods of dealing with them. Smear the top 10in (25cm) or so of each bud or bloom-carrying stem with a light coating of petroleum jelly. It may sound difficult, but it is really quite simple. Rub some of the jelly onto your hands and then smooth down each stem as you move along the rows of plants. The

Spraying has to be undertaken regularly during the growing season.

warm summer air will disperse the mixture along each stem without harming the blooms at all. No earwig, however active, can pass this barrier, and your buds and blooms will be safe from damage. Caterpillars too are deterred by this jelly barrier, which is an extremely popular way of safeguarding blooms. However, caterpillars do attack in other ways if you are not careful. The main offender is the flitting, cabbage-white butterfly, which seems to lay as many eggs on dahlias as it does on its natural host plants, the brassicas. The eggs are usually laid on the undersides of leaves, and the caterpillars will eat your plants to shreds if you ignore them. If the "dancing whites" are in your garden, it is time to douse all your plants extra well with a malathion-based insecticide.

Insect pests, as great a nuisance as they can be, are not to be confused with diseases. They can cause disease as we have already seen, but disease attacks on dahlias are usually fatal. Although infected plants may linger in your garden, they really should be removed and burned to avoid the possibility of the healthy plants catching the infection. One such disease, and perhaps the only real menace, is *D. mosaic*. It is called mosaic because the symptoms on the leaves are a paling or yellowing of the veins until they look like a true mosaic. Other signs are more obvious: the leaves become misshapen and have blisters on the undersides, and the most unmistakable sign of all to help you identify mosaic virus is the fact that the plants resolutely refuse to grow more than 1ft (30cm) high, which has given *D. mosaic* its common name, "stunt virus". If you have an attack and it reveals itself in any of the symptoms described above, you will be doing yourself and all the dahlia growers in your immediate vicinity a favour by uprooting the offender and burning it immediately.

Lifting your Tubers

As with all good things, the dahlia season comes to an end. And in October (sometimes in November) with the arrival of winter, the first strong frost will blacken the foliage of your plants, and you will be faced with the job of saving those precious tubers, which have been forming underground as you admired and picked the flowers above.

The question is always asked: "Is it necessary to lift dahlia roots?" The answer is that, even if you live in favoured areas of the United Kingdom like Devon and Cornwall, it *is* necessary. Many growers living in the south-west would agree, and although you might well be lucky and survive a series of winters, there will come one that will take every tuber that you leave in the garden and reduce them to useless pulp. At least if you extract and store the tubers, you will have a chance of getting pleasure from them next summer.

Tuber lifting and storage starts with the removal of the old foliage. Cut down the stems to a few inches and clear the whole plot, adding all the leaves to your compost heap. If you are fussy about names, move the label to the shortened

At the season's end prepare to lift the tubers by clearing away canes, mulch and other debris.

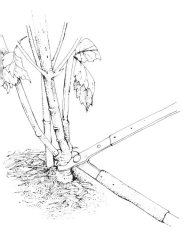

Cut down the old, tough stems with secateurs.

stem, twisting wire firmly around the stem so that the label is not lost during the storage period. Use a spade rather than a fork, which can damage the fat roots by "spearing" them, to lift the root and lift carefully to avoid snapping off portions at the point where they join the stem. A lot of soil will come up with the tuber – they are often very large – so probe away as much of the surplus soil as you can, using a blunt stick (a plant label is ideal) and making sure that you do not mark or damage the surface of the tuber. Take the tubers into your shed or greenhouse and allow them to dry out naturally. Doors and windows should be left open on good days (close them at night if frost is forecast) so that by the end of a week or so, a vigorous shake will remove the remainder of the garden soil that had been clinging between the tubers. It is at this point that many gardeners would store their roots, but I still have a few jobs to recommend you do before you commit your tubers to their winter hibernation.

Preparing roots for storage is, in my view, the right way to get most of them safely through even the worst winter. The main danger is an attack from fungus, with frost running this a close second. Reduce the chance of fungus attack by shortening the stem to only an inch or so. Then make a drainage hole in the soft centre of the stem by thrusting a screwdriver or keyhole saw through it and out at the bottom. Next, trim off the smaller, hair-like roots and even up the thong-like pieces to about $\frac{1}{4}$in (6mm) in diameter, leaving only the tight, compact tuber itself.

There remains just one more task to do before packing away the now easy-to-handle dahlias, and that is a fungicide treatment. Nowadays dahlia lovers choose the new chemical benomyl (sold in sachets as Benlate) and if you make up

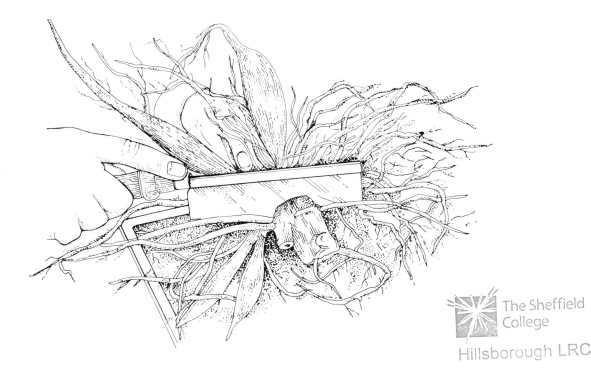

Remove the old stem, cutting it to an inch or so in length, to make storage easier.

The Sheffield College
Hillsborough LRC

31

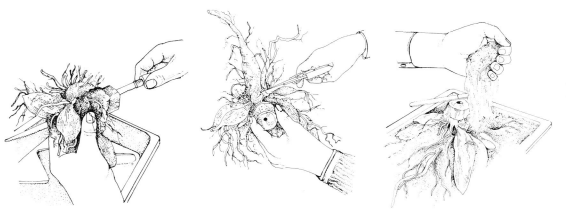

Make a hole right through the old stem (far left) by thrusting a keyhole saw through its soft centre, trim off all the hair-like roots (centre) to leave a compact tuber for storage and pack away the prepared roots in shallow trays (left) covering them with dry soil or peat. Don't forget to inspect them regularly.

a 2-gallon (9-litre) bucket, it is an easy matter to dip the roots into the solution. Allow them to drain and dry for a few hours before storing them.

For many years my own dahlia tubers have survived the rigours of the winter packed in shallow trays and bedded completely into a mixture of dry soil, peat and sand. Set the roots carefully on a bed of the mixture, then cover them so that the entire root is obscured but the shortened stem just breaks the surface. The advantage of using shallow trays lies in the fact that you can examine the contents regularly without disturbing the tubers too much. If you see any signs of fungus attack, remove the offending plant and destroy it at once!

Your completed store trays should now be placed in a frost-free place. A greenhouse where a modicum of heat is kept on is perfect: I store mine on the staging so that the soil-warming cables keep the frost away. A garden shed is fine, but always have sacks or some other heavy material available to cover the trays in really bad conditions. Many gardeners use their garages, and if they are brick-built, they are usually frostproof, which cannot always be said of the fabricated types. The secret is to make sure that frost cannot get to your stock, but wherever you choose to keep your tubers, it is necessary to make regular inspections throughout the storage period at least once a fortnight.

If you take these few simple precautions, there is no reason at all why you should not get the bulk of your dahlia tubers safely through the winter. Don't be disappointed when you lose some of your roots. Even the most venerated and experienced of dahlia growers will freely admit that it is virtually impossible to achieve 100 per cent success rates. It has been done, but so rarely that ancient gardeners tell of such deeds in hushed voices around the autumn bonfires!

And so the cycle is complete and another season starts. The intricacies of dahlia cultivation may, at first, seem confusing. But don't be deterred; the end product of these pleasant tasks is the bringing to life of one of the most beautiful flowers in the world today. And what better reward for a little devotion can you truly expect?

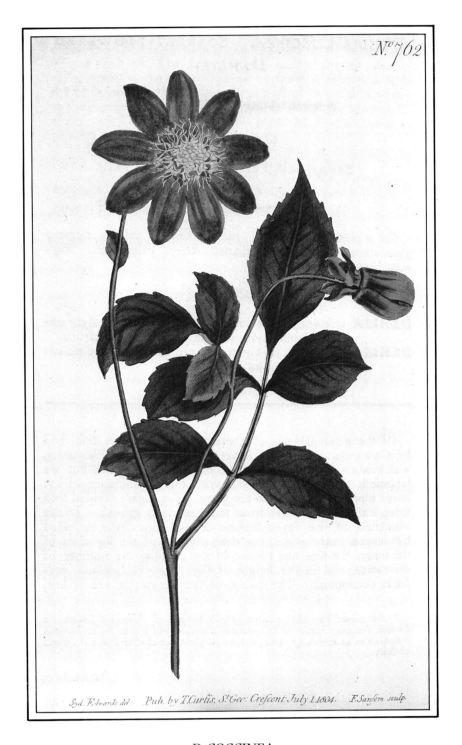

Syd. Edwards del.   Pub. by T.Curtis, St Geo: Crescent July 1.1804.   F.Sansom sculp.

D. COCCINEA
One of the founders of the modern dahlia dynasty,
D. coccinea is a natural hybrid, which
originated in Mexico and central America, where it was
discovered by Francisco Hernandez, a Spanish
botanist almost 200 years ago.

SUPER TROUPER

It seems that everyone loves the combination of white to lavender/lilac, as the lists contain dozens of varieties with this mixture. I loved the elegant stems and fine form of ''Super Trouper'' when I first grew it two or three years ago, and there is no doubt it will be extremely popular when better known.

JESCOT LYCIA

A legacy of that great dahlia raiser of the 1950s and 1960s, Ernest Cooper from St Albans in Hertfordshire, "Jescot Lycia" is, in the words of many enthusiasts, many years ahead of its time. This strap petalled form could not find a home on the National Dahlia Society register because it did not fit any of the existing types. It was promptly called a double-flowered orchid variety and placed in the miscellaneous section. There is no doubt that we will see more of this form in the future for it serves well as a cutflower and as a subject for flower arrangement.

BETTY BOWEN

This truly versatile dahlia may be used for exhibition, garden decoration or cut-flower work. Who says so? No other authority than the National Dahlia Society itself, which rates this Yorkshire dahlia very highly indeed. A tall grower – some 5ft (1.5m) or more – the bright purple blooms have the regular (formal) make up that is so beloved by showmen. Its height makes it ideal for the back of a border, and the strong stems make it useful when you are looking to fill vases.

DANA DIANA
"Dana Diana" first took my eye when I saw it on trial at Wisley. It eventually received a Highly Commended, but had the decision been mine, it would have received a medal! I love the delicate bronze shades, and the semi-cactus form, halfway between the broader decoratives and the spiky cactus, adds to its charm. A fine dahlia, it would be a bonus in any collection.

HAMARI BRIDE
This white medium, semi-cactus variety from dahlia hybridist "Pi" Ensum, was so named when two of his daughters were married in a single ceremony on the same day. Every top honour has been given to this one, which remains a banker for show even though it is over twenty years old.

REGINALD KEENE
"Reginald Keene" first saw
light of day in Suffolk. It is
from the inspired camel-
hair brush of Geoff Flood of
Beccles, who has prefixed
many of his successful
raisings with the word
"Suffolk". "Reginald
Keene" was an exception,
but it is probably his best
ever. A large semi-cactus,
in blends of orange and
flame, it wins readily all
over the dahlia world and
tops popularity charts year
after year.

SO DAINTY

This miniature semi-cactus has delicate bronze colouring, and it is so neatly constructed that it is bound to catch the judge's eye! It is recommended by the National Dahlia Society for exhibition as well as for gardeners in search of varieties suitable for cut-flower work or for their borders. Like the miniature (pure) cactus types, it is in a very rare group.

HAMARI KATRINA

Although "Hamari Katrina" rates as a leading exhibition variety, when it is grown naturally – that is, without too much attention to disbudding – it will grace your garden all summer long with its elegant pale yellow blooms. Semi-cactus in form, it is larger than the average garden dahlia being 8–9in (20–23cm) in diameter. A large vase, however, and a ready cutting of "Katrina" and your friends will be full of admiration for your skills!

KAY HELEN

Good new poms are very rare indeed, and good new poms from Britain are even rarer, as most originate overseas. This beauty, in shades of pink and white, was not only raised in Hampshire, England (by Charles Burton), but it took a major award, the A.M., at Wisley in 1981, following this with the top First Class Certificate in 1984.

PORCELAIN

With its sculpted waterlily form, "Porcelain" is one of the dahlia finds of the 1980s, and it received the highest award at Wisley, the coveted First Class Certificate. As the name would suggest, the colour is a startling translucent blend of white and lavender, and the varying shades make it perfect for both borders and cut-flower work. Raised in Sussex by John Crutchfield, a leading professional until his recent retirement, "Porcelain" is a dahlia *par excellence*, helping to prove decisively that the world's best dahlias are British!

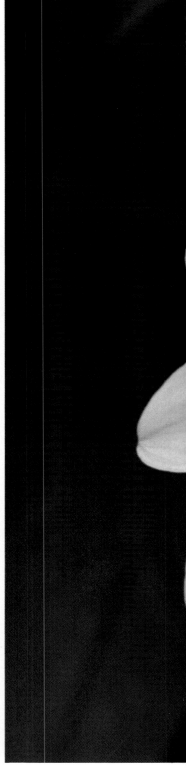

GIRAFFE & SON
It is almost forty years
since the rare variety
"Giraffe" (above left)
appeared at the Hague
nurseries of Van Hoek.
Since then, only one "son"
– or, to be more specific,
one sport – "Pink Giraffe"
has appeared. The name
was not because of
the tallness of the plants
on which the variety
blooms, but because of the

strange bronze and yellow markings, which somewhat resemble the colour of giraffes. Although it is officially known as a double-flowered orchid form, it is placed in the "miscellaneous" section because there are not enough of the double-flowered orchid form dahlias to make a separate section.

PURBECK LYDIA
The attraction of this beautiful cactus dahlia is its flamboyant colour. The flame-red blooms put it ahead of many of its less well endowed rivals. It has strong growth and exhibition potential and is rated extremely highly. It was raised near Weymouth, Dorset, by one of the newer British hybridists, Bill Cann.

MAJESTIC ATHALIE
This well-formed cactus in blends of dark pink and yellow is a sport of *Athalie*. A sport from an established parent is of value if the showman knows and has mastered the requirements of the parent; it follows that the colour sport will require only a repeat performance to emulate its parent.

DE LA HAYE

"De la Haye" is a new dahlia from Britain. That isn't really news these days, as Britain just about leads the world in the production of new dahlias, especially those that are ideal for exhibition. "De la Haye" is just such a dahlia in the medium cactus group. It won a major seedling class at the National Dahlia Show in London only a couple of years ago, and, as it is also a prolific form, you have a perfect product!

RUSTIG

We get many fine dahlias from overseas, especially from western Europe and the United States; this pale yellow medium decorative comes, however, from a lesser known dahlia-loving country—South Africa. The different cultural conditions of the country of origin often deny a 100 per cent performance by a foreign raising, but "Rustig" has proved to be an exception to that rule.

WISLEY BEDDING TRIALS
The bedding Trials, which are held at Wisley, Surrey, the home of the joint National Dahlia Society and Royal Horticultural Society trials, are always a joy to behold, spreading a sea of colour over the whole garden. Bedding types can be raised from seed, and nowadays there is a whole range of forms, colours and heights to satisfy even the most demanding gardeners. Bedding dahlias are one of the professional parks gardener's main allies in his annual battle to fill our public parks with colour each summer.

PINK JUPITER

Where, you might ask, is "Jupiter" if we have a pink form? The answer is that the parent of "Pink Jupiter" was the top show variety "Daleko Jupiter", a blended red and yellow giant semi-cactus. This rich pink sport is a true offspring of a winning mother. Honoured at Wisley – not many giants have that privilege – it goes from strength to strength on the show bench. In 1985 it had most winning cards at the Harrogate (N.D.S.) Show and was beaten by one card at the London Dahlia Show by – "Daleko Jupiter"!

NEVERIC
This fine dahlia from Yorkshire is in the large section. It is a semi-cactus and a blend of bright orange and flame—a real eyecatcher. Another to be awarded at Wisley, the powerful stems and excellent form make it a fine candidate for exhibition, and it has proved itself on many occasions! Neville Weekes from Morley, Leeds, raised this one.

RED ALERT
Another Trials Winner, with a very worthy Highly Commended to its credit in the first year, "Red Alert" comes from nurseryman Philip Tivey, who, with his son Christopher, trades internationally from a nursery at Syston near Leicester. A compact miniature, so beloved by most gardeners, this scarlet beauty has beguiled many enthusiasts, who now grow it to the exclusion of some of the older red miniatures. Long stems make it perfect for flower arrangers!

CATHERINE IRELAND
This elegant and cuttable miniature decorative was raised in Surrey only a few years ago, and yet it has achieved great popularity. A winner at the National Dahlia Society and Royal Horticultural Society's Wisley Trials, you can show this one or use it for floral arrangements.

BONNE ESPERANCE

This dark pink, open centred Lilliput dahlia is from a large family of tiny beauties that have a useful role to play in any planned garden. Growing only to 12in (30cm) or so in height, they need no staking and require only that the dead or dying bloom heads are removed to ensure a continuous flowering from July until the first frost of winter. Whether on a patio or edging a border they will perform unaided for your pleasure. You will find them under the title Lilliput, Topmix or, most often, Baby dahlias, in your nurseryman's catalogue.

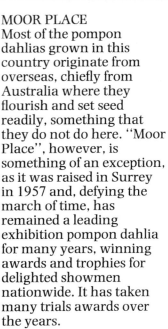

MOOR PLACE

Most of the pompon dahlias grown in this country originate from overseas, chiefly from Australia where they flourish and set seed readily, something that they do not do here. "Moor Place", however, is something of an exception, as it was raised in Surrey in 1957 and, defying the march of time, has remained a leading exhibition pompon dahlia for many years, winning awards and trophies for delighted showmen nationwide. It has taken many trials awards over the years.

HALLMARK

A pompon *par excellence*, the National Dahlia Society rates "Hallmark" officially (in its *Directory*) as good for exhibition, garden use and cutflower – a trio of values that are not attributed to many dahlia varieties. A blend of several pinks, even a lilac flush on occasions, it would have to be described as dark pink, its predominant hue. Originating from Australia, the home of the best pompons, "Hallmark" is a front-ranking exhibition variety, and no self-respecting dahlia showman would be without this winner.

HAMARI DELICE
The delicate form and colour, so much admired in the modern dahlia, are seen here to perfection. White to lavender, "Hamari Delice" is hardly unique, for many varieties have this colour combination, but add the reflexing form of the miniature decorative and the firm stem of the natural cut flower and the result is a dahlia to treasure. The prefix "Hamari", which apparently means *Chez nous* in Hindustani, is the hallmark of hybridist "Pi" Ensum, and, as might be expected, "Hamari Delice" has already received Wisley Trials honours.

SCARLET COMET
There are very few dahlias of the anemone-flowered type, and the bright red "Scarlet Comet" is a sport of one of the originals, "Comet", which is a much darker red. Raised in East Anglia some twenty years ago by nurseryman Cyril Piper, "Scarlet Comet" has never reached the heights of popularity of some forms, but it will fascinate visitors to your garden who will almost certainly doubt that it is a dahlia at all!

LEILA MARY

Although it has the fascinating form of the collerette, "Leila Mary" has the unusual combination of white on white, whereas this group is more usually grown for its flamboyance of colour, with bi-colours and variegated types abounding. The clear white of both the inner and outer petalling contrasts quite beautifully with the bright, golden yellow of the open seedhead. The French raised the first collerettes around the turn of the century, and this thin-stemmed and prolific group is a favourite with both flower arrangers and those just longing to fill a vase with elegant blooms.

LILAC JESCOT JULIE

The miscellaneous forms, like this double orchid-flowered, are extremely popular with those who practise the art of floral arrangement. Like its parent "Jescot Julie", it is from the late Ernest Cooper, who specialized in this type of bloom and gave us so many winners to cherish.

VUURVOGEL
One of the rarest colours in commerce today, this dahlia, with its unusual blend of scarlet and bright yellow, is found in nurseries all over the world, where it is often named as "Firebird", the English translation of its apt Dutch name. The Netherlands dahlia traders, who sell some 50 million tubers each year, were quick to see that such a brightly coloured dahlia would be attractive to gardeners, and they have included this beauty in their highly organized marketing schemes that ensure that dahlias can be bought at supermarkets and chain stores all over the world. Usually priced at less than £1.00 for a tuber, this popular beauty has to be a "best buy".

HOLLAND FESTIVAL
This giant decorative dahlia (above right) can easily be grown to 12in (30cm) in diameter. The combination of bright orange and white tips is unique, and no other dahlia in the same form and size group can match it. It is grown worldwide for both show and spectacular display. Raised in Holland as you might expect, it came from the highly specialized nurseries of Firma Bruidegom at Baarn. It is a quarter century old now, and took an Award of Merit at the Wisley Trials in 1963.

LA CIERVA

Since it was first spotted in its native Holland some 36 years ago, "La Cierva" has been a faithful servant of dahlia lovers. This brightly hued, purple-tipped, white open-centred bloom, that falls into the little known Collerette group, has been winning prizes at dahlia shows for most of that time, and if enthusiasts were asked to name one variety that they grew in this section, they would nominate "La Cierva".

CHERIDA

The beautiful colour combination, a subtle blend of what is (officially) called bronze and lilac shades, is the main reason for this variety's popularity. A miniature ball type, which means that the mature bloom is about the size of a tennis ball, the stems give it power in a vase, and it is a favourite for flower arrangers at any level of skill.

JESCOT JULIE

Listed by the National Dahlia Society as a miscellaneous variety, "Jescot Julie" is double orchid-flowered, one of only three such varieties classified. The strap-like petals in shades of dark pink and bronze make it look very much like an orchid. It was raised, like all the "Jescots", by the late, great dahliaman, Ernest Cooper.

RHEINFALL
The skilful nurseryman
Cornelius Geerlings of
Heemstede, Holland, sends
many of his new dahlias to
Britain for trial at Wisley
Gardens, and "Rheinfall"
received a Highly
Commended award in its
first year there. A cut-
flower type, which will
give plenty of blooms from
July to the first frosts, this
brilliant small red cactus
on excellent stems will
satisfy even the most
pernickety gardener!

ROKESLEY MINI
This white beauty is one of a very rare section indeed. There are only four miniature cactus varieties – officially so classified – and this is one of them. It was raised in London and awarded at the Wisley Trials.

HAMARI SHIMMER
From the world's premier dahlia hybridist, "Pi" Ensum, "Hamari Shimmer" lives up to the very high standards that this skilful raiser asks of his seedlings. Another in the favourite white to lavender/lilac shades, it has already been commended at Wisley, and more awards seem likely to follow. Cut and vased with other pink or soft cream varieties, it is an absolute stunner.

PEACE PACT

Another waterlily-formed variety, this brilliant white form is from Holland, where it received the highest trials awards. "Peace Pact" was awarded at Wisley and went straight onto everyone's "wanted" list. It will stand proudly to be admired in your border, may be cut with relish and will win prizes at your local show in the class for waterlily-flowered dahlias. What more could any dahlia lover want?

SYMBOL

Doubly awarded at Wisley Trials (highly commended in 1962 and winner of the Award of Merit in 1966), "Symbol" still wins regularly and is regarded by many as the standard all dahlia raisers should seek to equal. A medium semi-cactus, this bronze winner was raised by Holland's top dahlia expert, Dirk Maarse of Baarn, the owner of the famous Bruidegom Nurseries now, unhappily, no longer trading.

HAMARI GOLD
Ace dahlia raiser "Pi" Ensum, who won the national title on many occasions, knows a winning dahlia when he sees it, and "Hamari Gold", a bronze giant decorative with impeccable form, is one of his most recent successes. It went straight to the hearts of dahlia lovers worldwide, and in a few short years was a firm favourite from Birmingham to Brisbane! A low-growing variety, it can achieve a diameter of 12in (30cm) with perfect ease.

INCA DAMBUSTER
Perhaps the largest giant variety ever raised in Britain, this pale yellow semi-cactus can be grown to diameters of 16in (41cm). George Brookes, one of the dahlia world's senior hybridists who has been responsible for many hundreds of successful introductions and who has probably received more awards at the Wisley Trials than any other raiser, gave us this super dahlia.

POLYAND
We get many fine dahlias from Australia, and this elegant lavender large decorative is no exception. It has dominated the large decorative group for many years and seems destined to do so for many more. Easy to grow to perfection, it is a dahlia that may be recommended to the beginner, for the form and size are assured. You may see it referred to as "Polyanna", but "Polyand" is the official registration.

L'ANCRESSE

Dahlias are named by their originators after many things—favourite aunts, uncles or even pets. This lovely white miniature ball dahlia from Surrey is named after the raiser's favourite holiday spot— L'Ancresse in the Channel Islands. A national trophy winner as a seedling, it is now a leading show variety.

HAMARI ACCORD

This newcomer has yet to be classified by the National Dahlia Society, but I would anticipate that it will be a large semi-cactus. Bright yellow, it is yet another in the long string of world-beating dahlias from "Pi" Ensum, and it was a winner of the nation's top trophy (as a seedling) when released two years ago. This is a dahlia with a future, and one that you can grow with confidence.

AMBER BANKER
Orange is the true colour of this sport of the famous Dutch dahlia "Banker", which is bright red. The pure cactus forms are not as popular for show work as the semi-cactus varieties, but a few, and this is one, compare very favourably with their more powerful brothers!

ALLTAMI DANDY
This elegant small decorative in pastel pink and yellow took the Award of Merit at the Wisley Trials in 1984. It obviously found favour with the Trials judges, and it is not hard to see why.

D. IMPERIALIS

D. Imperialis or tree dahlia as it is often called,
is the tallest of all the species varieties.
Growing to heights of 20ft (6m) or more, it is of little
value as the flowers are insignificant. The
Mexican Indians used the hollow-stems as waterpipes.

THE SHOW BENCH

The dahlia showman, it has to be said, is not the rare bird that so many believe it to be. Hundreds of gardeners, both male and female, devote most of their spare time to the production of superb dahlias, and they are joined in a truly international fellowship that has branches reaching from Australia, to Asia, South Africa, the Americas and into Europe.

But what constitutes a good dahlia? And what makes one dahlia better than another? Beauty may be in the eye of the beholder and there are some differences of opinion throughout the international community, but there is agreement on most of the basic criteria: cleanliness, form, size and colour. So rather than detracting from the whole, the differences that arise within the various organizations tend to be stimulating rather than confusing.

The organizations that are responsible for deciding just what a good exhibition dahlia should look like are the national dahlia societies, of which the largest, organizing the world's biggest dahlia show every September, is Britain's National Dahlia Society. Based in London, it works closely, most of the time, with the Royal Horticultural Society. Operating through its special committees, the council of the National Dahlia Society frames the rules for judging dahlias and circulates them to the Society's 4,000 members to keep them abreast of developments or any new rules or groups of cultivars. All dahlia cultivars are classified in the Society's very comprehensive *Classified Directory and Rules for Judging* (see page 94), which is printed bi-annually. The Classified Directory is available not only to the Society's U.K. members; it is sold internationally, because, despite the existence of their national ruling bodies, some overseas societies choose to have their shows judged by British rules (and give British awards) rather than use their own. Although this "poaching" is on a minor scale, it does tend to stimulate the overseas societies!

But how do these N.D.S. committees decide what a dahlia should look like? On what do they base their decisions? Experience, of course, has a lot to do with it, and in the thirty years that I have lived with and loved this flower and its attendant administrative bodies (I have been General Secretary of the N.D.S. for twenty years), I have reached the conclusion that dedication and devotion play equally great parts. Although there might be some who would criticize the decisions that the committees have taken, the overwhelming opinion of dahlia lovers is that the rules are right. And so they should be after more than one hundred years, because the N.D.S. was formed on Christmas Eve 1881.

If you are contemplating a foray into the world of dahlia exhibiting, let me assure you immediately that there is very little dividing the ordinary dahlia in the average dahlia grower's garden from those that will bring a silver cup or two to your sideboard. Many of the varieties that are grown to be used as cut flowers or for garden decoration are exactly the same varieties that can and do win top prizes. The difference lies only in the cultivation, which does require just a little more attention.

The world of show dahlias is dominated by size. Grow a dahlia over the sectional size limit and your exhibit – in any group except the giants, which you may grow as big as you are able – will be disqualified. This size rule is very strictly interpreted by the 400 qualified dahlia judges on the National Dahlia Society's official listings. Each judge carries a set of size rings, which are dropped over the blooms to ensure that they do not exceed the size limits. If you were thinking that such a rule would be likely to create a little controversy, you would be right, and showmen have never stopped arguing about the rights and wrongs of the system since it was introduced some ten years ago. The use of the rings is an official rule, however, even though it can cause some heartache. I vividly recall judging in France at the Golden Jubilee show of the French Dahlia Society. As is my wont, I took my rings and waved them, probably unconvincingly, over some of the French growers' blooms. My Gallic friends were hysterical. "You measure dahlias in England, Philippe?" they asked in amazement. I had to admit that we did, and I have no doubt that they put it all down to our eccentric national character.

If you are still set on showing dahlias, therefore, the first thing you must do is to learn the sectional size limits and, of course, buy a set of rings. If your dahlias are not to be disqualified, incurring the dreaded N.A.S. – Not According to Schedule – these are the size limits imposed by the N.D.S.:

Pompons must not exceed 52mm (2in)
Miniatures must not exceed 115mm ($4\frac{1}{2}$in)
Small must not exceed 170mm ($6\frac{3}{4}$in)
Medium must not exceed 220mm ($8\frac{3}{4}$in)
Large must not exceed 260mm ($10\frac{1}{4}$in)
Giant-flowered varieties may be allowed to grow unchecked

Growing a dahlia to a particular size may seem a daunting task. In reality, however, it is not, because the popular show dahlias will grow naturally to the limits specified or, to be more accurate, to within a few centimetres of the limit. Of course, the object is to create a bloom that teeters on the edge of the specified size, that is, in effect, as large as possible within the allowed limit. Experience and experiment are the keys to achieving this. You will be able to adjust the size, up or down, by leaving on the plant sufficient blooms to keep them inside the rings or by taking away excess growth to encourage them to grow. This may sound complicated at first, but once you have got the hang of it, debranching (taking away some of the flowering stems *before* they bloom) and disbudding (to help increase size) come naturally. Below is a guide to the approximate number of stems that you should leave on your plants for the various sections:

Pompons: allow the plants to grow naturally without any restriction; some disbudding may be needed near to show time
Miniatures: do not exceed 12 stems
Small: no more than 12 stems on strong growing plants
Medium: keep to 8–10 stems per plant
Large: restrict to 5 or 6 stems per plant

You may by now be thinking that exhibitors spend most of their time avoiding too large blooms, and that indeed has to be uppermost in their minds, but other aspects of show flowers are equally important, and the word that is used to describe such aspects is "ideal", a word that is only too familiar to dahlia exhibitors. It is these "ideals" on which the framework of perfection hangs.

For instance, blooms must be symmetrical in all respects, and the outline of the flowers must be perfectly circular. The depth of the dahlia, that is, the distance between the front of the bloom and the furthermost petals at the back, should be at least two-thirds of the width, more if possible. The centre has to be in proportion to the rest of the flower, and there must be no gaps in the outline of the petals. The angle of the bloom, that is, the way the bloom sits on the stem, should not be less than 45°; this applies to all types except the pompons, which must, "ideally", sit full square on top of the stem. The stems must be perfectly straight and in proportion to the size of the bloom. I have never understood what stems have to do with bloom perfection, but it is something else that exhibitors must contend with.

In addition to all these factors, the "natural" things that contribute to the elegant and exceptional dahlias that could be termed "show" dahlias are also paramount. If you place a set of dahlias in a vase for the attention of the judge, each dahlia must be clean and undamaged by anything from bruising to the marks of insect bites. A wilting flower, or anything that departs from the formation described above or a flower that has an open centre (the so-called "daisy-eyed") will get short shrift from the officials, and in every show where N.D.S. judges officiate their word is unchallengeable.

So, how are you to ensure that your dahlias meet all these criteria? As I mentioned earlier, there is little difference between dahlias grown for everyday purposes and those grown for exhibition. It is simply a question of giving your dahlias more attention and of doing today what needs to be done today, not leaving it all until the weekend! Obviously, every exhibitor has a set of guidelines that he follows to keep his blooms in front. Here is my list:

1. Start with the right varieties. It is no good believing that your lovely dahlias will win just because *you* like them. Study the winning charts from the shows and grow those varieties that do well in your area.

2. Always obtain your stock from top-class sources. Tubers or plants bought from general suppliers are normally not the best; get to know the top suppliers and buy from them.

3. Always, but always, grow exhibition dahlias on their own. A plot or bed properly spaced and laid out will put you on the right track immediately.

4. Consider covering the blooms, especially if you want to grow the large or giant types. Polythene covers stretched over the plot keep the maturing blooms clean and damage-free, which is half the battle.

5. Before you enter a large show, go as a visitor to study just what the exhibitors are doing. You will be surprised to find how ready and willing they are to chat about their techniques with you.

6. Try and buy your own staging equipment. Although vases are supplied by some of the bigger societies, it is best to have your own. Make sure that your containers are the correct size, and never try to place an exhibit in the wrong vase – large blooms in a small vase spell disaster.

7. Join your local specialist society; there is one in every area of Britain, from Scotland to the West Country. As you progress, join the National Dahlia Society, and you will soon find its annual analyses and other literature invaluable. The Society will help you keep abreast of the best varieties – those that are winning prizes – and generally keep you informed of everything pertaining to the shows, like dates, new venues, trials results, etc.

The attractions and rewards of dahlia exhibiting are many, but let me say at once that hard cash is not one of them. Except in very exceptional circumstances, you will never make enough prize money even to cover your expenses. The cost of stock, fertilizers, insecticides, potting soils, canes, travelling and entry fees to shows all add up dramatically, and the most that you can expect in return is a few pounds, and that only if you win, of course. But what you will have is the pleasure of achievement, and if you have never known the thrill of seeing a first prize card propped against your exhibit, then you have something very special in store. And add to this the joy of the friendship you will find as a member of the exhibiting fraternity. I have made hundreds of friends over the years all over the world, and that is something far more precious than any monetary reward.

THE FUTURE

The dahlia is still a flower of mystery. Hidden behind its beautiful face and winning smile is a host of wonderful things that have yet to be revealed to the eager gardener. Some of these will emerge soon, others not perhaps for another hundred years. We have already mentioned the quest for a blue dahlia, a dahlia with a beguiling scent and a dahlia that will resist frost. All these should eventually be possible, but in my view they are just the tip of the iceberg.

Many dahlia enthusiasts believe that the creation of a dahlia type that will set double-flowered seed will be the ultimate achievement. Imagine being able to save seed from your dahlia plants every autumn safe in the knowledge that by sowing these in the following spring, you will be able to re-create the fully double scarlet cactus or the white pompon from which the seed was saved. The implications of such a step would be that there would no longer be a need to lift and store tubers; propagation need never start until late spring; and the production of tubers for sale by the trade would be a thing of the past. Perhaps nurserymen would not be overjoyed, though! Experiments are being conducted at the moment, and each year the possibility that we will have fine quality seeds producing consistent double flowers becomes more of a reality.

Then there is that long-forgotten dahlia project – the creation of a food source. There are scientists who fervently believe that the dahlia could be encouraged to carry a mass of elegant flowers during the summer months and then, when lifted, offer itself as a tasty root crop. In the 19th century such experiments failed dismally as we have seen, but we are almost in the 21st century now, and it cannot be beyond the skills of scientists to enhance the flavour of the tubers to make them at least reasonably palatable. Why no one has ever tried I cannot understand. The potato, carrot and the turnip have all been subjected to the closest scrutiny at national vegetable research stations, and, it has to be said, scientists have achieved considerable success in terms of better yields, earlier cropping and so forth. There is no reason why the dahlia should not offer the same opportunities for experiment and perhaps with the same results – a combination of beauty and food.

Another area for experimentation is the development of dahlias that flower earlier or later in the year. The creation of a frost-resistant strain of dahlia has long been an ambition of serious dahlia lovers, but some growers offer an alternative to creating a frost-proof plant: rather they suggest that the dahlia could be made to flower twice in one season, say a main flush in early summer and a further full flush in mid-summer. Years ago in the United States, experiments were carried out with a growth activator known as giberellic acid, which worked on corn crops. When it was applied to dahlias, the plants raced into bloom, without apparent ill-effect, in half the normal time. Unfortunately, perhaps because the novelty wore off or for some other unknown reason no more has been heard of it. Could giberellic acid be the key? Could this ingredient, culled from the mould found on rice seed, be the catalyst that brings us the frost-free or double-season blooms?

My own dream is that one day a British scientist will walk into the Royal Horticultural Society in mid-December carrying a bunch of blue dahlias, a scent of roses or lilac wafting around them. Dreams, I must remind you, do have a habit of coming true.

RECOMMENDED VARIETIES

In any list of recommendations – especially of flowers –
there is bound to be a personal element. Many of the dahlias listed here
are varieties that I love and have grown well.

Giants and Large-flowered Dahlias

Giants and large-flowered dahlias are particularly suited to show work; although some gardeners do grow them to enhance their gardens and to cut for the house, their true destiny is the show bench.

Alvas Supreme
A pale yellow from New Zealand, decorative in form; the finest show giant we have; ideal for the beginner

Daleko Jupiter
A very big semi-cactus in blends of bright red and yellow; an established winner

Hamari Girl
One of the easiest giant decoratives to grow to show perfection; a rich pink; good for the beginner

Hamari Gold
One of the finest of the post-war British raisings; colour as name; low growing at around 3ft (1m)

Polyand
An Australian import, but superb on the show bench; a large decorative with lilac hues on excellent stems

Reginald Keene
A leading large semi-cactus in orange flame blends; a beautiful British-raised show variety

Cut-flower Varieties

Cheerio
An unusual cherry red tipped with silver semi-cactus; a cut flower *par excellence*

Doris Day
Dutch, despite the name; a dark red cactus on long stems; the perfect cut flower with a great reputation

Downham Royal
An enviable, rich, royal purple colour; the fashionable ball type; will cut and show

Gerrie Hoek
A pink waterlily; the best known of all dahlias for cut-flower work; still going strong after forty years

Glorie Van Heemstede
A lovely butter-yellow waterlily form; a Wisley winner in 1986 despite being forty years old

Hamari Fiesta
A British-raised decorative in a rare yellow-tipped scarlet; a universal favourite for garden and cut-flower work

Klankstad Kerkrade
A classic dahlia: a sulphur yellow cactus for cutting and a proven show winner

Lady Linda
A top show bench winner that cuts for the home beautifully; small, yellow and reflexing to the stem; a personal favourite

Match
A vivid white with purple tips from South Africa; very popular for both cut flower and floral art

Nina Chester
An elegant white to lavender decorative; will also double for show work

Rokesley Mini
A true miniature cactus in white; first-class cut flower; Wisley awarded

Wootton Cupid
A pink ball and so versatile; scores of blooms to cut and a prize winner too

Pompons

Perfect for floral art work, and as subjects for your own arrangements in the home pompons are unbeatable. Remember that for show work they need to be grown no larger than 52mm (2in) in diameter.

Hallmark
A beauty; a pink that is both floriferous and show worthy; my personal favourite

Moor Place
A deep purple, perhaps the best of the purple pompons; powerful stems and good for show work too

Pop Willo
Best described as corn coloured, this globular beauty is another for show or vase filling

Small World
Almost in a class of its own; a rare white pompon that is another dual-purpose variety – i.e., show and cut flower

William John Newbery
A very recent British introduction that has impressed greatly, a deep red for show or cut flower; awarded at Wisley

Willo's Surprise
A dark red pompon from Australia; excellent for garden and show

Six Dahlias that are Different

Comet
This Australian-raised anemone-flowered beauty in dark red is a show stopper; grow it in your garden and all your friends will want some of your stock

Giraffe
Officially a double-orchid flowered type that resembles that exotic species; a mix of yellow and bronze that is stunning in a floral arrangement; try also **Pink Giraffe** the sport, colour as name

Jescot Julie
Another personal favourite; described as double-orchid flowered although the strap-like petals are a new form; orange petalling is backed with purple – breathtaking

Mariposa (Butterfly)
A collerette-flowered dahlia in dark lilac pink with a startling white collar; one of the best

Princess Marie Jose
A single-flowered dahlia in lilac with a bright yellow eye

Nurseries

Dahlias may be purchased from suppliers in one of three forms: as tubers, as rooted cuttings or as plants. Tubers, which are usually available in the popular pot-root form (i.e., a plant that has been grown through one summer in a small pot), should be ordered during the late autumn or early new year, when they can be used for the taking of cuttings (see pages 21–3).

Rooted cuttings are ready in late spring, and they will be sent to you quickly by ordinary post. It is then up to you to pot them on and care for them (in your greenhouse or cold-frame) until they are ready to go out in the open garden.

Plants, raised and cared for by the nurserymen until they are dispatched to you, can be ordered so that they are delivered at the correct time for you to plant them in the open garden. This will be late spring onwards for gardeners in the south of the country, with those growers in the colder north having to wait a week or two more before risking their young plants in the open.

Write to your supplier to ask for a catalogue; also make sure you ask about the conditions of sale. Only a few professionals will, for example, send out plants by post or rail; most prefer you to collect in order to prevent the damage or delays that might result from road or rail transport. Tubers and rooted cuttings are not, of course, subject to the same restrictions, and they are usually dealt with through the postal services.

If you order from overseas, you must ensure that you have the necessary import permissions. To make sure that you are complying with the current regulations, write or telephone to the local office of the Ministry of Agriculture; you will find the address and telephone number in your local telephone directory. Although there is no charge for obtaining permission to import tubers, you do have to complete certain forms to qualify. Some overseas countries, such as Holland and Australia, require the package to be covered by a health certificate, so make sure you order from a recognized supplier who knows *all* the regulations. Remember: you may not import plants of any sort.

United Kingdom

Aylett Nurseries Ltd, North Orbital Road (A405), St Albans, Hertfordshire (Telephone: 0727 22255)
Perhaps the largest dahlia nursery in the country; specializes in cut-flower and floral art types; does not send out plants; watch for special "open" days in September

Braintris Dahlia Nursery, Beccles, Suffolk NR34 7RL (Telephone: 0502 715728 or 715489)
An exhibition variety specialist but with a good list of garden dahlias; exports all over the world and carries many novelties

Bebbington's Dahlias, Lady Gate Nursery, 47 The Green, Diseworth, near Derby DE7 2QN (Telephone: 0332 811565)
All types available; the best imported from overseas, propagated and offered for sale to U.K. growers; plants, tubers or rooted cuttings sent out

Halls of Heddon, West Heddon Nurseries, Heddon on the Wall, Newcastle upon Tyne NE15 0JS (Telephone: 06614 2445)
A very extensive list containing something for every aspect from exhibition giants to bedding types; plants dispatched by road/rail

Butterfields Dahlia Nursery, Harvest Hill, Upper Bourne End, Buckinghamshire (Telephone: 06285 25455)
Some of the rarer types available (e.g., Topmix, collerettes and a range for floral artists); welcomes individual visits or groups at flowering time by arrangement; does not dispatch plants by road/rail

Oscrofts Dahlias, Sprotborough Road, Doncaster, Yorkshire (Telephone: 0302 785026)
An extensive list with novelties and some own raisings; tubers or plants and special offers of "own choice" selections that can be real money savers

Les Staite & Sons, Avon Nurseries, Evesham, Worcestershire (Telephone: 0386 6212)
Bulk supplier to the trade, but would cater for groups or societies; personal visits would assist individual purchases

Dobie & Son Ltd, Upper Dee Mills, Llangollen, Clwyd LL20 8SD (Telephone: 0978 860119)
This famous firm has a good range of dahlia seed; a wide range of tubers is also available

Suttons Ltd, Hele Road, Torquay, Devon TQ2 7QJ (Telephone: 0803 62011)
Supplies an excellent range of seed dahlias as well as tubers of varieties suitable for cut-flower or exhibition work; imports from overseas also, mainly Holland

On the Continent

Bloemisterij G. Aartsen, Vlierburgweg 51, 3849 MB Hierden (bij Harderwijk), Holland (Telephone: 03410 12089)
A superb range of dahlias for the ordinary gardener, the exhibitor or even the professional; tubers (no plants) by post or personal collection; most new continental raisings carried; visits in the flowering season by arrangement

Lindhout Ornata, Speciale Dahliaculturen, Herenweg 105, 2201, AE Noordwijk, Holland (Telephone: 01719 15567)
Carries a lengthy list of all types, including the best of Dutch and other continental introductions; tubers sent in season

C. Geerlings (Dahlien Culturen), Kadijk 38, 2104 AA Heemstede, Holland (no telephone number)
A dahlia trader with many varieties of his own raising that have won honours all over Europe; tubers only available and all prices quoted in Dutch guilders

Wilhelm Schwieters (Dahlien), 4427 Legden (Westf.), West Germany (Telephone: 02566 1233)
Much awarded professional, with a comprehensive list of varieties for garden and cut-flower work mostly of German origin

Otto Bergerhoff (Dahlien), 5276 Wiehl 1 (Bezirk Köln), West Germany (Telephone: 02262 93113)
One of the best known "dahlia names" in Germany; a professional raiser with a quality list to match

United States

When you order stock from the United States, bear in mind that a tuber as it is known in Britain is not the same in America. If you ask for a tuber you will probably receive a portion or "split" of a tuber. European enthusiasts have christened these "chicken legs", because they resemble that particular delicacy.

Comstock Dahlia Gardens, P.O. Box 608, Solana Beach, California 92075
Perhaps the most famous dahlia professionals in the U.S., with a long list for all purposes including some of their own raising; a bonus – sun-ripened seed at very reasonable prices

White Dahlia Gardens, 2480 SE Creighton Ave., Milwaukee, Oregon 97222
A long-standing firm with a fine reputation

The Blue Dahlia Gardens, San Jose, Illinois 62682
Owned and managed by Ken Furrer, one of the best known dahlia people in the States, who travels all over the world to add to his massive list, including visits to Holland, Belgium and Germany

The Phil Traff Dahlia Gardens, 1316 – 132nd Avenue East, Sumner, Washington 98390
A recent arrival on the professional scene, offers a mix of international raisings with the emphasis, naturally, on those raised in the U.S.A.

Almand Dahlia Gardens, 2541 W. Avenue 133, San Leandro, California 94577
Jack Almand is a leading raiser and trader with many fine introductions to his credit; tubers or seeds are available

Czechoslovakia

JZD MIR – Turany (Dahlias), Zahradnictvi 050, 627 00, Brno-Slatina
A small but effective list with some new Czechoslovak novelties as well as imports from all over the world

Australia

John & Anne Menzel (Dahlias), Box 27, Berri 5343, S.A. A very large list including many rarer types such as the anemone, stellar, pincer (incurves) and orchid types

Japan

Yusaku Konishi (Dahlias), 455 Chibadera, Chibashi
The lists are in Japanese, with many of them overprinted in English; tubers come well from Japan, but will be expensive as they *must* be air-freighted

I would fervently recommend air freight for the import of all dahlia tubers from overseas, apart from Europe, of course.

GARDENS TO VISIT

Dahlia displays, whether they be in the form of shows, trials or garden displays, can be seen all over the world whenever the season is in full spate. In Europe, that means during the months of August, September and, quite often, October.

Dahlia growers in Britain are fortunate in that the two best dahlia shows in the world are held in Britain, and certainly the best trial garden is British. The main event is the *National Dahlia Society's London Show*, which is held every year in early September at the Royal Horticultural Society's Halls, Westminster, London S.W.1. Here you will see everything that is good in the dahlia world, including the latest varieties from overseas and the newest winners of the current seedling classes. The national dahlia exhibitors' championships are held here each year, some ten of them, and they alone are worth the entrance money.

The *N.D.S. Harrogate Show* is considered to be the second best all-dahlia show in the world (London, of course, being the best), and this is staged in north Yorkshire during the third week of September. Again, all that is best will be on view, with the added bonus of a stunning display by the trade. In 1986 the Dutch trade erected a superb 80-foot stand with all the best of the continental introductions on display. The show is currently held in the main Exhibition Centre.

The *Wisley Trials* are generally thought to be the very best dahlia trials in the world. They are a joint venture by the Royal Horticultural Society and the National Dahlia Society, and they attract thousands of visitors every summer. The awards, made during the autumn, are accepted worldwide as a hallmark of the success of a new dahlia. Wisley is in Surrey on the A3, near Ripley, and admission is free to members of the R.H.S., although a charge of £1.25p is made for others. It is open to members only on Sunday mornings.

Bradford Exhibition Trials, the National Dahlia Society's exhibition trials, are held in Brackenhill Park, Bradford, West Yorkshire, every summer, and, despite the off-putting name, they are well worth a visit, because they are always a blaze of colour and contain some of the finest varieties raised today. Admission is free, and the garden is open every day. Results of the trials are usually known by mid-September.

Dahlia enthusiasts have voted Roger Aylett's Dahlia Festival weekends a great success, and they are certainly well worth a visit. The *Aylett Gardens and Dahlia Festivals* are located on the North Orbital Road (A405) near St Albans, Hertfordshire. The plantings are extensive and on two consecutive Sundays at the end of September a festive atmosphere reigns as competitions are held, floral art is demonstrated and the whole gardens open for gardeners to explore. More details can be had from Aylett's Dahlias, North Orbital Road (A405), St Albans, Hertfordshire. All of the nurseries mentioned on pages 89–90 are open to the general public, and if you require advice on varieties or help with solving a particular cultivation problem, the dahlia professional is your man. If you do intend to visit, it is advisable to telephone or write to find out when the nurseries are open.

The Sheffield
College

Hillsborough LRC

On the Continent

One of the largest dahlia displays-cum-trials is held every year on the island of *Mainau*, in Lake Constance, Southern Germany. It really is massive, and thousands of plants are on view. Dutch, German, Belgian, Austrian and Italian professionals, among others, visit Mainau, which offers lavish awards to the successful. As is to be expected, a large planting like this attracts large numbers of visitors, and if you are holidaying in the area and want to see the show, remember to avoid weekends.

There are several trials and show areas, often held in the Cologne area of Germany, but for obvious reasons they are moved around the country. Write to Elisabeth Goring, Ubierstrasse 30, 5300 Bonn (Bad Godesburg), West Germany (Telephone: 0228 362057) for the very latest information.

Holland is described by many as the home of the European dahlia and it is true that the flower is a way of life for many. The export of between 50 and 60 million tubers every year is an important slice of the country's economy. There are plenty of places to see the dahlia on trial or in shows, but, as in Germany, they are moved around the country so that the gardeners of the Netherlands have a chance to see just what their professionals can achieve. For up-to-date information about shows and trials in Holland, write or telephone Peter Lindhout, Secretary, Netherlands Dahlia Society, Lindhout Ornata, Herenweg 105, 22ol AE Noordwijk, Holland (Telephone: 01719 15567).

In addition, I personally recommend that you visit any of the nurseries mentioned on page 90. All will offer you a very friendly welcome.

In *France* the dahlia organization is somewhat fragmented. Thriving groups operate in the north, centre and in the south-west at Biarritz. You can get information from Georges Clenet, Secretary, Société Française du Dahlia, "Champs Elysees", Rue de Baudreuil, 02100 St Quentin, France. Occasionally the society's trials are held in the grounds of the Cathedral at St Quentin – a beautiful if unusual site for such a display.

Denmark has a very active dahlia organization, but for the latest show and trials location you will need to contact the secretary, whose full address is: Dansk Dahlia Selskab, Andreas Bjerggård, DEG's, Konsulent Kontoor, Post Box 3073, 1508 Copenhagen V, Denmark.

There is a thriving dahlia society in *Belgium*, but it is almost wholly professional. Nevertheless, it does hold some superb shows and trials, details of which will be revealed if you write to the society's secretary: Marcel Martin, Struikenlei 13, B2120 Schoten, Belgium. These shows and trials are often combined with displays of begonias for which, of course, Belgium is famous all over the world.

Other national organizations exist worldwide, here are just a few for your information. *Australian Dahlia Society:* G. Parker Esq., 9 Acacia Avenue, Dernancourt, South Australia, 5075, Australia. *American Dahlia Society:* Mrs Irene Owen, Secretary, 345 Merritt Avenue, Bergenfield, N.J. 07621, U.S.A. *National Dahlia Society of South Africa*: J. Walker Esq., Secretary, "The Shieling", 3 Turvey Street, Port Elizabeth, South Africa. *National Dahlia Society of New Zealand*: Mrs P. Hawkins, Secretary, 8 Sharyn Place, Tauranga, New Zealand.

USEFUL INFORMATION

The National Dahlia Society can be contacted through

The General Secretary
National Dahlia Society
9a High Street
Kingsthorpe
Northampton
NN2 6QF

Ask for an enrolment form and look out for the Society's special offers, usually in the autumn, when it gives two years' membership for the price of one subscription.

The N.D.S. runs a free advice service (please enclose a stamped, addressed envelope) for members and non-members, which will locate any variety you need. Sometimes the addresses are overseas, and you can obtain the stock (as tubers) without too much trouble if the addresses are within the continent of Europe (see page 90). Special conditions apply, however, to suppliers outside what is called the "Mediterranean area", and you should ask your local Ministry of Agriculture for more information.

Apart from the National Dahlia Society's London Show and the Harrogate Show, scores of other specialist shows are held nationwide, and you can obtain information on venues and dates currently in operation from the Society (please enclose that essential stamped, addressed envelope). Also available from the Society are the *Classified Directory and Rules for Judging*, sets of exhibitors' rings and a cultural guide.

INDEX

Bold numerals refer to captions
to colour illustrations